Civil Engineering Construction

B.G. Fletcher
Head of Department of Construction, Waltham Forest College

S. A. Lavan
Senior Lecturer in Civil Engineering, Westminster College

Heinemann : London

William Heinemann Ltd
10 Upper Grosvenor Street, London W1X 9PA

LONDON   MELBOURNE
JOHANNESBURG   AUCKLAND

First published 1987
© B. G. Fletcher and S. A. Lavan 1987

British Library Cataloguing in Publication Data

Fletcher, B.G.
   Civil engineering construction.
   1. Civil engineering
   I. Title II. Lavan, S.A.
   624   TA145

  ISBN 0 434 90572 0

Publishing services by Ponting-Green, London and Basingstoke
Photoset by Parker Typesetting Service, Leicester
Printed in Great Britain by
M & A Thomson Litho Ltd, East Kilbride

Contents

Preface

The main purpose of this textbook is to provide civil engineering technicians with an introduction to the principles and practice of civil engineering construction. The authors have kept in mind during its preparation the fact that at technician level the availability of textbooks dealing exclusively with civil engineering rather than building construction is limited. In terms of the level of treatment of the subject matter the authors believe it should serve well students studying BTEC ordinary and higher certificate and diploma courses at levels III and IV as well as a primer for undergraduates following degree level courses in civil engineering.

A textbook of this nature cannot claim to provide a comprehensive guide to all aspects of the work of a civil engineering technician – few, if any, could. It is, therefore, essential for a successful understanding of one's work that the student should augment his or her studies by reading the industry's magazines and newspapers and, of course, if passing a site stop and look. Lecturers and textbooks, important as they are, form only part of the process of understanding the ways and needs of one's chosen career.

The authors would like to extend their thanks to David Wallis for preparing the chapters on Falseworks and External Works, Trevor Fairman who has so expertly prepared the illustrations and Felicity Hunt who so patiently typed the manuscript. Last, but by no means least, those employed in the industry who, when asked for advice and assistance in obtaining material, were quite splendid in their response.

B.G. Fletcher
S.A. Lavan

1 Contract administration

Introduction

A civil engineering project is usually initiated by a promoter, who may be an individual, but is usually a public utility, local authority, government department or a large private company.

The promoter will know the type of project he wishes to undertake, such as a power station, refinery, reservoir or road complex. At an early stage he will enlist the services of a civil engineer to advise on the design of the structure(s), planning permission (if necessary) costing the project and the selection of a contractor.

A large proportion of civil engineering work, particularly small or medium projects are handled by the promoter's own staff. This is useful if the work is of a specialist nature, as the staff will have a detailed knowledge of the promoter's requirements.

The principal advantages of retaining a consulting engineer are that his judgement and advice are independent of all outside influences and that he is a specialist, spending most of the time on the design and construction of new works.

A civil engineering contractor will undertake the construction of the works, to the requirement of the engineer, for a given sum of money. There are various methods of selecting a contractor for a given task and these are discussed later.

A civil engineering project may, therefore, be considered to be brought to a successful conclusion by the combined skilled efforts of three main parties – the promoter, the engineer and the contractor.

Types of contract

There are several types of contract which can be used for civil engineering work. These include:

Bill of quantities – this is probably the best type of contract. It incorporates a bill of quantities which includes individual items, lump sums and provisional sums. The bill is priced by the contractor and a sum can be arrived at for the construction works.

It is advisable to use this form of contract for most civil engineering jobs, as the method allows all contractors to price the work on exactly the same basis and gives the contractor a very clear idea of the work to be carried out.

Lump sum – if the project is not too large and full details are available (drawings and specification), a contractor may be able to offer a single fixed lump sum price for the works.

The advantage of this form of contract is that detailed measuring work is avoided and the promoter has the assurance of a fixed total price. There can be a disadvantage if the engineer wishes to change the design once construction work has started or unforeseen circumstances occur.

Schedule of rates – this type of contract is useful for maintenance work. The usual method is for a schedule to be prepared and the contractor to insert a rate against each item of work. The contract is fair to the contractor but does not give the same confidence as a bill of quantities contract.

Target – the contractor has an incentive to complete the work as quickly and economically as possible. With this type of contract the method is to quote a basic fee as a percentage of an agreed target estimate based on the bill of quantities. The fee can fluctuate depending on the final estimated cost. If the design is modified in any way, it will be necessary to adjust the target figure. This may result in a dispute between the engineer and contractor if excessive changes have to be made.

Package deal – in this case the promoter will approach a contractor with a broad outline of his requirements. The contractor will submit full design details, method of construction and costing of the project. This type of contract is only suitable for large multi-million pound projects such as oil refineries, power stations etc.

Other types of contract are available, but these are usually a variation of the above.

Contract documents

Various documents are required in connection with a contract and these are listed below. It will depend on the type of project whether all of them are required. Documents include:

Contract Drawings
Bill of Quantities
Specification
Form of Tender
Form of Agreement
General Conditions of Contract

Contract Drawings – these show details of construction and, in addition, the nature of the site, access, and a full report on ground conditions should be made available to a contractor.

Bill of Quantities – this consists of all the items of work to be included in the contract. Quantities are usually measured in accordance with the Standard Method of Measurement for Civil Engineering. The contractor is required to state a rate for each item of work, which should include an allowance for overheads and profit.

Specification – describes the work to be executed in detail and supports the contract drawings by providing further information on the nature and quality of materials and workmanship.

Form of Tender – is the formal offer by the contractor to complete the contract for an agreed sum and includes the time scale for the completion of the project. The form in general use is included in the Institution of Civil Engineers Conditions of Contract.

Form of Agreement – the contractor enters into an agreement with the promoter, who is now termed the employer, to construct the works for an agreed sum and to conform in all respects with the provisions of the contract.

General Conditions of Contract – the conditions are outlined in the *ICE Conditions of Contract* referred to previously and specify the terms which the work is to be undertaken. It also states the powers of the engineer and how and when payment by the employer is made and the contractor's general responsibilities for all site operations.

Letting the contract

The normal procedure is for the promoter to advertise by way of notices in the technical press that he is open to receive tenders for a certain job. The advertisement will give a brief description of the works and the location.

As an alternative, it is sometimes necessary for the engineer to prepare a list of selected contractors who are invited to submit a price for a job. This arrangement is used if the project is of a specialist nature.

In some cases, particularly on very large projects, the engineer may negotiate with a single contractor to undertake the construction works.

Whichever method is employed the engineer will undertake a close scrutiny of the tenders submitted and will possibly discuss the offers with the contractors before recommending to the promoter which offer to accept. It should be noted that it is the promoter who finally decides if the offer is to be accepted.

The construction team

To obtain a clear impression of the personnel and their tasks involved on a civil engineering contract, it is necessary to distinguish between the contractor's and engineer's staff and employees.

When a contract has been placed it will be necessary for the contractor to arrange a construction team who will be responsible for the organisation and running of the contract.

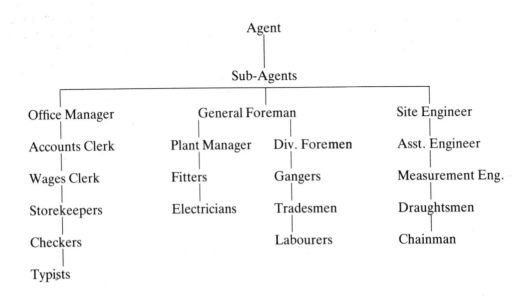

Figure 1.1   The contractor's site team

At the same time the engineer will appoint a resident engineer whose main role is to supervise the works on behalf of the promoter.

The contractor's site team

A typical team for a medium-sized project might be as shown in Figure 1.1. The key posts are:

(i)   agent
(ii) site engineer
(iii) plant engineer
(iv) general foreman
(v)   office manager

Briefly the responsibilities of the five posts are:

(i) The agent

The agent is responsible for the overall running of the contract and will have the authority to hire men and plant, purchase materials and employ sub-contractors.

He must, of course, have an extensive knowledge of civil engineering techniques, but need not necessarily be a chartered civil engineer.

Many agents have risen from the ranks of the craftsman but as the techniques of construction have become more sophisticated many younger agents today have academic engineering qualifications.

(ii) The site engineer

The site engineer is responsible for the setting out of work. This will include taking site levels, lining in and levelling constructional work, planning and designing temporary works such as access roads, concrete batching plant; foundations and dealing with power and water supplies and the drainage of the site. The engineer will also be required to act as adviser to the site agent and keep progress and quality records of the work.

(iii) The plant manager

His is a most important job as it is his responsibility to keep all the mechanical plant in working order and to have it available as required by the programme of construction.

(iv) The general foreman

The general foreman's task is to keep the contract running to programme. He will issue to the trades foreman the detailed instructions of work to be carried out and, if necessary, be able to show how it is to be done. To many people he is the lynch-pin of the contractor's team.

(v) The office manager

The office manager is responsible for the administration work such as issuing orders for materials making up pay sheets, receiving and checking accounts.

Generally though, many jobs on the site will overlap and chains of organisation will vary according to the practices of the contractor concerned and the type and size of contractor.

The contract manager

On large contracts the contractor will appoint a contract manager who, working from the head office will be concerned with the management rather than technical aspects of the contract. The manager in turn will be directly responsible to a director of the company.

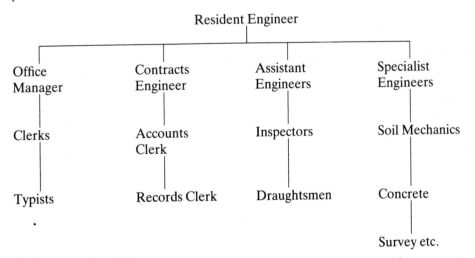

Figure 1.2   The resident engineer's site team

The resident engineer's team

The consultant engineer will appoint a resident engineer (not to be confused with the contractor's site engineer) who, according to the size of the contract, will have a number of assistants. Such a team, for a medium project, might be as shown in Figure 1.2.

The resident engineer's main responsibility is to see that the construction of the project is in accordance with the designs and drawings issued by the consultant engineer and any specialist consultant (e.g. mechanical or electrical engineers) employed on the project. This will include checking that material specifications, workmanship, levels and setting out are as required, to measure completed work for the purposes of interim payment and to calculate such payments.

To assist him in his work the resident engineer will have a staff of assistant engineers and inspectors (in building contracts the clerk of works is the equivalent post). The number depends on the size and nature of the contract.

It is an important part of the resident engineer's duties to keep records of the work carried out. These records should enable a clear view of the progress of the contract at any time to be obtained and form the basis on which as mentioned above, interim payments will be made to the contract. The engineer should also record the quality and performance of the materials used and the actual dimensions etc., of the completed work in the form of 'as built' drawings.

Contract planning

Before a contractor can start work he must plan how the site is to be laid out. Of equal importance is the planning of the order of construction. Without such a plan the contract would soon run into difficulties and is therefore considered a standard procedure.

The complexity of tasks involved on any one civil engineering contract will, of course, vary according to the type of work to be carried out, from the simplest single pipe laying contract for a local authority to major schemes such as a power station or a dry dock complex. Whatever the contract though, the contractor must be able to state how long it will take to complete, how many men will be required at any time during the contract and what materials and in what quantities they will be required at any time in the contract.

When the contractor has decided upon the order of construction he will submit his scheme to the consultant engineers for approval. If approved he will then start on a detailed plan of the work to be carried out. There are several methods used for planning, the main difference between them being the level of sophistication. The basic requirement of any plan, and its presentation, is that it will show in a manner that is clear to all who have to read it, the expected starting and completion dates of the individual jobs, that make up the project. Linked with this is the number of skilled and unskilled operatives required and the quantity of materials to be used. This last point is very important as the late delivery of materials can hold up the contract and this can prove to be very expensive for the contractor. By planning his material requirements beforehand, the contractor can obtain delivery dates and if necessary alter his plan to fit the available delivery periods.

The plan is usually referred to as a 'programme of work' and is normally presented in the form of a bar-chart. At first appearance a bar-chart may seem simple enough, but one should not be deceived. A great deal of work and experience must go into it if it is to be a realistic programme that can be adhered to during the contract.

To avoid producing a bar-chart programme that is difficult to read due to the amount of information contained in it, it is more usual to produce several bar-charts. This is normally done by having a bar-chart that will show the periods of time (and in some cases the materials required) for the major items of work for the complete contract. Augmenting the overall bar-chart are weekly or monthly bar-charts showing in greater detail the work to be carried out in any particular week or month. Each bar-chart should be marked up as the work proceeds so that the progress of the contract can be assessed by referring to the bar-charts. Figure 1.3 shows a typical bar-chart.

Monitoring of progress

To ensure the smooth and effective progress of work on a construction site, good planning is of paramount importance.

In order that the programme of work is adhered to, a regular review of the progress of operations is essential. Progress records are usually maintained by the general foreman, but in the case of a large contract, a work study or progress engineer is employed to ensure that the delivery and placing of materials is maintained and that labour is available to ensure that the programme is kept to schedule. To check progress, a measured weekly survey of works is carried out and each operation compared with the programme.

When there is a considerable variation from the programmes, it may be desirable to adjust the planned sequence of operations. In such circumstances the general foreman or progress engineer would consult with the contractor's planning department before making such changes. Close co-operation between the site staff and the contractor's planning department is usually achieved by means of site meetings when it is

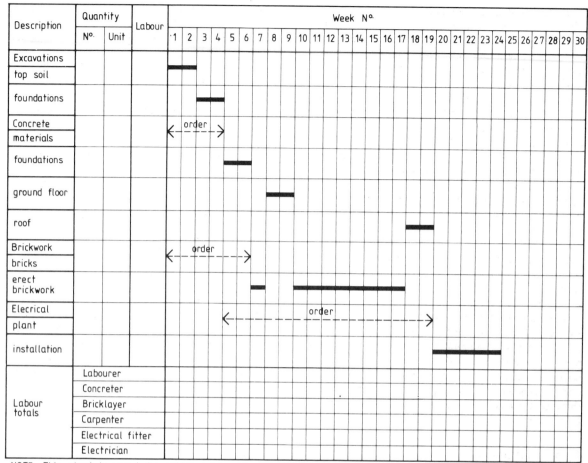

Figure 1.3 Part of a bar-chart for an electricity substation

usually possible to correct any discrepancies in the approved planning programme.

The site staff maintain a record of current and future strength of the labour force and also keep careful records of all materials delivered and any balance outstanding.

In addition, a regular meeting of all interested parties (consulting engineer, quantity surveyor, contractor and, if necessary, the architect) is held to discuss progress of the contract and, if necessary, action can be taken to correct problems associated with the progress of the project.

Statutory requirements

A contractor, as does any employer in the United Kingdom, has a legal obligation to ensure the health, safety and welfare of his employees at their place of work. This duty is enshrined in four sets of Regulations under the Construction Regulations 1961. They are:

1 Construction (General Provisions) Regulations 1966
2 Construction (Lifting Operations) Regulations 1961
3 Construction (Working Places) Regulations 1966
4 Construction (Health and Welfare) Regulations 1961

These Regulations apply to any type of building work and most types of civil engineering and construction work.

1 General provisions
This Regulation provides for safe working in the following areas:

(i)   Excavation, shafts and tunnels
(ii) Cofferdams and caissons
(iii) Explosives
(iv) Dangerous or unhealthy atmospheres
(v)   Work in or adjacent to water
(vi) Site transport
(vii) Demolition

2 Lifting operations

This Regulation covers the use and maintenance (by required regular inspections) of lifting appliances (i.e. cranes and derricks) chains, ropes, lifting tackle and hoists. It also covers the carriage of persons and the secureness of loads.

3 Working places

This Regulation provides that a safe route must be provided to and from every place of work on the site and that every place of work must also be safe. Any work that cannot be done from the ground or part of the structure must have scaffolding erected for working from. It is this Regulation that covers the requirements for scaffold structures.

4 Health and Welfare

This Regulation, since its introduction, has had certain parts amended by the introduction of the Construction (Health and Welfare) (Amendment) Regulations 1974. The Regulation provides for the general health and welfare of construction site operatives and includes the provision of facilities for dealing with accidents, shelter, messrooms, washing and drying out facilities and sanitary arrangements.

It is planned that with the introduction of the Health and Safety at Work Act 1974 all existing safety legislation will eventually be replaced by this Act. Under this Act every employer of more than five employees must prepare a written statement of the company's health and safety policy.

There are other acts and regulations which relate to certain aspects of civil engineering work such as the Mines and Quarries Act 1954; Diving Operations Special Regulations 1960 etc.

Under the Construction Regulations an employer who employs more than twenty persons must appoint a safety officer. This position may not necessarily be a full time one, but whoever is appointed must be given sufficient time from any other duties he has, to carry out his safety duties with reasonable efficiency. His name must be entered on the copy of the Regulations that the employer is required to display on site, normally in the contractor's office.

2 Preliminary works

Soil investigations

The understanding of the physical makeup, engineering and chemical properties of the sub-soil strata of a site is of fundamental importance to the safe design of the structure. To obtain this information a soil investigation must be carried out.

Two codes of practice are of particular relevance to this work and reference to them for a full understanding of the requirements and techniques involved is essential. They are:

(i) BS5930 (1981) Site Investigations;
(ii) BS1377 (1975) Methods of testing soils for Civil Engineering purposes.

The term sub-soil is used to describe the layers of strata that lie between the top-soil (see Figure 2.1) and the bedrock. It is these layers of soil that in most cases support or surround (in the case of tunnels for example) a structure.

The scope of a soil investigation in terms of both extent and detail will depend on several factors. These may in general terms be stated as:

(i) the type and importance of the structure;
(ii) the existence of any previous knowledge of the sub-soil;
(iii) the cost of the investigation relative to the cost of the proposed structure.

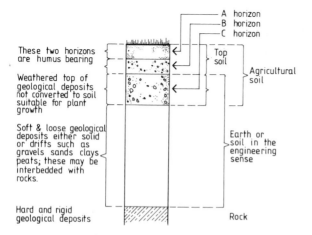

Figure 2.1 Relationship between engineering and non-engineering soils

Assuming then that a site requires a fairly detailed report, the soil investigation report should contain sufficiently detailed information on:

(a) the nature and thickness of the sub-soil strata;
(b) the mechanical properties of each strata or at least the strata and any underlying strata that a structure is to be supported on. Such information would include the density, natural moisture content (including any expected seasonal variations) compressibility, bearing capacity and shear strength;
(c) the chemical and physical properties;
(d) the natural ground water level of the site, again noting any seasonal variations.

This will enable conclusions to be drawn on the behaviour of the soil during excavation, construction and the working life of the structure. The report will also have a major influence on the type of foundation a structure will require.

Before samples of the soil may be taken, access must be made to the various levels of soil strata. The choice of how this will be done will depend on the nature of the ground (e.g. whether clay or running sand etc.), the topography of the ground and the comparative cost of the available methods.

The actual methods most commonly used are:

(i) Trial pits;
(ii) Borings;
(iii) Headings.

Trial pits

By digging a hole large enough for a man to work in, a section through the ground will be exposed, revealing for examination and sampling the soil strata.

This method is normally used for shallow exploration up to depths of about 3 m. Depths greater than 3 m do not tend to be economic in comparison to the second method.

Borings

As with a trial pit the purpose of a borehole is to reveal the make up of the underlying strata but, in this case, information is obtained by the collection of samples at suitable intervals as the borehole is sunk.

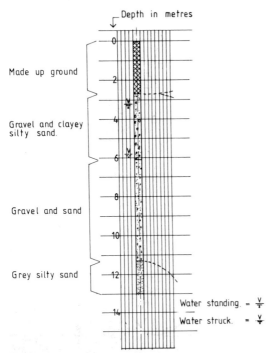

Depth in metres

Made up ground

Gravel and clayey silty sand.

Gravel and sand

Grey silty sand

Water standing. = ⩛

Water struck. = ⩛

Figure 2.2   Borehole log extract

A recommended layout of a borehole log is shown in Figure 2.2.

Boring Equipment

(i)   Post-hole auger

A simple hand operated tool that is suitable for depths of up to approximately 6 m in soft stratas (which will stand unsupported) is shown in Figure 2.3. In some cases such as gravels or loose sand the bore may be lined but the placing of the lining may require mechanical assistance.

(ii) Shell and auger boring

This tool can be hand operated in soft soils up to a depth of about 20 m with a diameter of 200 mm. For greater depths, a mechanical tool would be employed. Casing (if required) are positioned by means of a 'monkey' suspended from the winch – see Figure 2.4.

(iii) Percussion

As the name suggests the operation of a percussion borer is by repeated blows, breaking up the formation to sink the borehole. Water is added to the borehole as work proceeds, and the soil removed at intervals. Due to the action of the borer the collection of undisturbed samples must be carried out with care, and for detailed investigations the rotary method may be more preferable. An advantage of the percussion borer is its rapidity of progress.

Borehole logs

These are fully described in BS5930 (1981) but may be summarised as follows:

The borehole logs should give a clear picture in diagrams and words of the ground profiles at the particular point where the hole was bored. Most organisations carrying out site investigations have standard forms for recording information. It is seldom practical in these to make allowances for all data which possibly may need to be recorded.

Therefore, all logs are a compromise between what is desirable to record and what can be accommodated.

The following should be recorded on all logs:

(a)   title of investigation
(b)   report number
(c)   name of client
(d)   location detailed by a grid reference
(e)   date of boring
(f)   borehole number
(g)   type of boring
(h)   type of plant used
(i)   ground level
(j)   diameter of borehole
(k)   diameter of casing and depth to which taken
(l)   depth scale, such that the depth of sampling, tests and change in strata can be readily determined
(m)   depth of termination of borehole

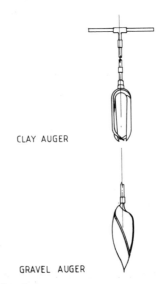

CLAY AUGER

GRAVEL AUGER

Figure 2.3   Post-hole augers

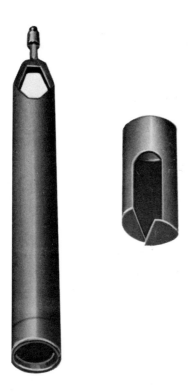

Figure 2.4   Shell with auger attachment

Figure 2.5   Typical mobile multipurpose drilling rig

(iv) Wash boring

With this method a tube is sunk by means of a strong jet of water. The jet of water disintegrates the soil and returns the disturbed soil by way of the returning current of water. Progress is made by either the tube sinking under its own weight, or driven by a 'monkey'.

(v) Rotary (Figure 2.5)

Two types of drilling are used:

Mud-rotary drilling
Core drilling

Mud-rotary drilling

In this system the bore is sunk by a drilling action with a rotating bit. As the hole is sunk a mud-laden fluid is injected into the borehole through the hollow rods connecting the bit to the rig.

In a similar manner to the wash boring technique, the fluid carries the disturbed soil from the borehole to the surface. The mud-laden fluid also acts as a support to the sides of the hole so casing is not necessary. Samples are obtained by use of a core cutting tool.

Core drilling

Used in rock, the core drilling technique produces a continuous core of rock. The broken rock displaced by the core cutter is removed in a similar manner as the wash boring method.

Borehole casings

In soft strata the borehole may collapse if the wall of the hole is not supported. The method most commonly used to prevent this happening is to line the hole with steel tubes, see Figure 2.6.

Figure 2.6   Typical borehole casing

Table 2.1   General guide to borehole diameters for varying depths

Total depth of borehole	Starting dia. of borehole	Final reduced dia. of borehole
Up to 20 m	150 mm	150 mm
20 m to 40 m	200 mm	150 mm
40 m to 60 m	250 mm	150 mm
60 m to 80 m	300 mm	150 mm

High tensile steel tubes in lengths of 1.5 m to 3 m with a male and female screwed ends, are driven down just ahead of the boring tool by a drop hammer operated by the same winch as the tool. Care must be taken when obtaining undisturbed samples or carrying out penetration tests at the base of the hole not to drive the casing too deeply, as this may affect the test results. The diameter of the casing should allow for a close fit boring tool.

To facilitate the withdrawal of the casing after the completion of the work in deep boreholes (i.e. over 20 m) the diameter should be reduced in stages. Suggested values for the reduction are set out in Table 2.1.

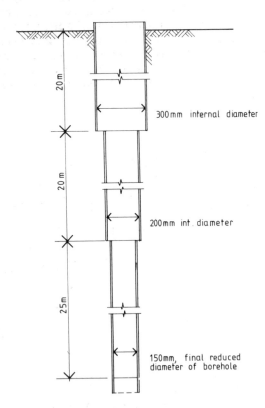

Figure 2.7 Reduction of borehole casing diameter to facilitate withdrawal

To illustrate the procedure, Figure 2.7 shows a typical lined borehole.

Headings

In suitable topographical conditions, headings driven either horizontally or inclined may be used. It is, for example, a particularly useful method to explore steeply dipping strata. In tunnelling work it is normal practice to drive a pilot tunnel ahead of the main tunnel to explore the ground conditions, this is a form of the 'heading' technique.

The choice of method may be summarised by reference to Table 2.2.

Insitu and laboratory testing of soils

Two types of soil samples are used for soil investigation analysis:

(i)   Undisturbed samples
(ii) Disturbed samples

Undisturbed samples

Certain soil tests require the sample of soil used to be in an undisturbed state. That is, the sample must retain the natural soil structure, moisture content and void-ratio of the soil from which it was taken.

Methods of obtaining undisturbed soil samples:

Trial pits

Hand samples of clay or sand are easily obtained from trial pits. For clay, a cube may be cut out with a sharp knife or a hand-corer of standard dimensions and known weight can be used (see Figure 2.8) The cylindrical cutter is pushed into the clay, pulled out and weighed. As the weight and dimensions of the cutter

Table 2.2   Soil investigation methods

Method/conditions	Soil	Rock
1. Trial pits	Open Timbered Piled Caissons	Open Timbered
2. Borings	Post-hole auger Shell and auger Wash boring Rotary	Percussion Rotary
3. Headings	Timbered Lined tunnels	Open Timbered

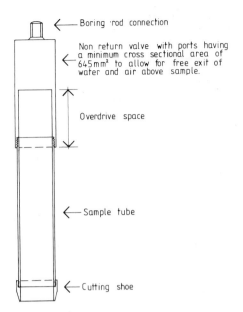

Boring rod connection

Non return valve with ports having a minimum cross sectional area of 645 mm² to allow for free exit of water and air above sample.

Overdrive space

Sample tube

Cutting shoe

Figure 2.8 Standard hand corer

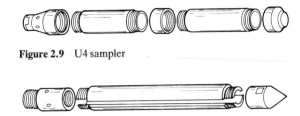

Figure 2.9 U4 sampler

Figure 2.10 Split spoon sampler

are known, the weight of the sample can be obtained by weighing the cutter and sample after the removal of the soil. Hence the soil density can be calculated.

The cylindrical cutter can also be used in comparatively moist sands but it is very difficult to obtain an undisturbed sample by this method in any other type of soil. A useful method for sand or gravel is to push into the soil a sampling tube or box open at both ends. The surrounding soil is dug away and trowel introduced underneath the tube or box. Level off the top and cover with a plate, and the sample can then be lifted out.

It is important not to take a sample from soil that has been exposed to the direct rays of the sun, wind or rain.

Boreholes

The core cutters illustrated in Figures 2.9 and 2.10 which are used for borehole samples should be well oiled both inside and out before attaching to the boring rods. This is to reduce friction, as frictional resistance whilst using the cutter may disturb the natural structure of the sample. The tool is lowered to the bottom of the borehole and forced into the clay by either jacking or by blows, though jacking is to be preferred. The distance that the core is driven into the soil should be carefully monitored to prevent the soil compressing in the cylinder. The non-return valve at the top of the sampler allows air to escape as the soil is forced up the cylinder. The tool consists of a cutting shoe, a sampler tube and a driving head. These three sections are detachable, allowing the sampler tube to be used as

container in which the sample can be sent to the laboratory. Figure 2.11 illustrates an undisturbed sampling technique.

To obtain undisturbed samples in water-laden soils a Bishop Sand Sampler can be used, see Figure 2.12. This is a compressed air sampler. The sample is removed from the ground into an air chamber and then lifted to the surface without contact with the water in the borehole.

When the undisturbed sample has been obtained it must be protected from damage and change in moisture content. As mentioned before the core-cutter tube can be used as a casing for transit to the laboratory, but the two ends must be sealed. The ends of the tubes should first be covered by waxed paper and then several layers of molten paraffin wax applied. Screw-on lids are then put on and sealed with tape. If the core is removed

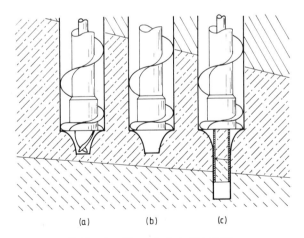

(a)   Auger flight and centre rod at desired depth.

(b)   Centre rod and drill bit withdrawn.

(c)   Sampling device introduced through hollow auger. Sections and sample withdrawn.

Figure 2.11   Hollow stem auger sampler

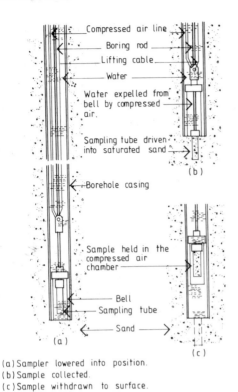

(a) Sampler lowered into position.
(b) Sample collected.
(c) Sample withdrawn to surface.

Figure 2.12   Bishop sand sampler

from the sampler, and this must be done with great care to avoid re-moulding of the sample, the sample should be wrapped in wax paper, coated in paraffin wax then placed in an airtight jar. The space around the sample should be well packed with sawdust to prevent damage during transit.

Disturbed samples

Because of the method of extraction, the natural moisture content and void ratio of a disturbed sample may vary from its insitu values, but the soil constituents will remain unaltered. Disturbed samples are used either for tests to supplement the information obtained from undisturbed samples or to determine the suitability of the soil for earthworks or stabilisation.

Such samples should be taken in duplicate, one for purposes of testing and one for the client, if required. For clays, at least 1 kg should be taken for each sample and placed in an airtight jar. Sands and gravels should be taken in greater quantities. The recommended minimum amounts are:

Medium grained soils – 5 kg
Coarse grained soils – 30 kg

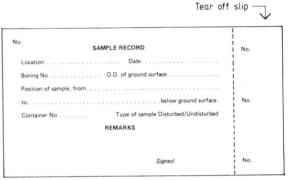

Figure 2.13   Recommended sample label

These samples can be stored in well tied polythene bags unless the moisture content is of importance, then the sample should be placed immediately on recovery in an airtight jar.

Because of the rapid variations that can take place in the properties of soils, clay in particular, and as these changes may not be apparent by visual examination alone, samples should be taken frequently. It is recommended that, if possible, a 100 mm diameter core be taken when a major change in soil type is evident and at about 1.5 m intervals during borings through apparently homogeneous stratum.

All samples should be labelled immediately after being taken from the borehole or trial pit. Figure 2.13 shows the recommended form to be used. These labels are pre-printed in duplicate and bound in book form. The serial number on the right hand detachable portion, which is attached to the sample, is printed three times so that the chance of it being defaced is diminished.

To summarise, the type of method of sampling, and the type of sample obtained are shown in Table 2.3.

It is frequently advantageous to determine the shear strength and density of the sub-soil insitu instead of obtaining samples from trial pits or boreholes for laboratory testing. Site testing is particularly useful in soft

Table 2.3   Summary of sampling methods

	Type of sample	*Method of sampling*
Soils	Disturbed	Hand samples Auger samples (e.g. in clay) Shell samples (e.g. in sands)
	Undisturbed	Hand samples Core samples

Table 2.4 Guide to density of soil according to number of blows

No. of blows/305 mm	Relative density
0 to 4	Very loose
4 to 10	Loose
10 to 30	Medium
30 to 50	Dense
Over 50	Very dense

clays and sands where boring may result in some disturbance of the soil structure making the recovery of undisturbed samples very difficult. Some of the tests used are described below:

Static penetration test

Sometimes referred to as the 'Dutch Cone' test due to its origin. The test consists of driving a rod with a point through a casing into the soil at a constant rate of loading. As the rod is cased there is no frictional resistance along the rod, therefore only the point of resistance is recorded, indicating the relative density along the penetrated depth. Only used in silty and soft strata conditions such as found in river estuaries.

Dynamic penetration test

Widely used in Britain and known as the Standard Penetration Test. In clays and sands a split-barrelled sampling tube of standard dimensions should be used – see Test 18 BS1377 (1975), and in gravels a closed conical ended rod.

Before the test commences, the base of the borehole should be relatively clean to avoid carrying out the test in partly disturbed conditions. The procedure is as follows.

The sampler is seated in the borehole by being driven 150 mm into the soil by means of a 65 kg drop hammer with a fall of 760 mm, the number of blows required to obtain the 150 mm penetration being recorded. The sampler is then driven 305 mm or until 50 blows have been applied by the drop hammer, again with a free fall of 760 mm. The number of blows required to obtain each 76 mm of penetration is recorded. The total number of blows required for the full depth of penetration, i.e. 305 mm, a record of the number of blows and depth of penetration is made.

Terzaghi and Peck recommend the following figures shown in Table 2.4 for the number of blows for full penetration as a guide to the insitu density.

Conclusions on the bearing capacity can be drawn from these results.

The California Bearing Ratio (CBR)

In the design of runways and roads the strength of the sub-grade is a principal factor in determining the thickness of the pavement. The strength of the sub-grade is assessed by the California Bearing Ratio of the soil.

For a general description of the apparatus and method of use, see Chapter 8. For a detailed description, reference should be made to BS1377 (1975) Test 16 and for guidance on the application of results, Road Research Laboratories publications *Road Note 29* should be consulted.

Vane tests

In soft clays it can be difficult to obtain reliable undisturbed samples to test for shear strength in a laboratory. To overcome this problem an insitu method of testing the shear strength of soft clays and silts is carried out. The test is known as the Vane Test and can either be hand worked or mechanically driven. Two types of apparatus are used, they are:

Borehole (Figure 2.14)
Penetration (Figure 2.14)

Basically a vane of cruciform section is driven into the soil and subjected to a torque force. The measure of shear strength is the relationship of the torque required

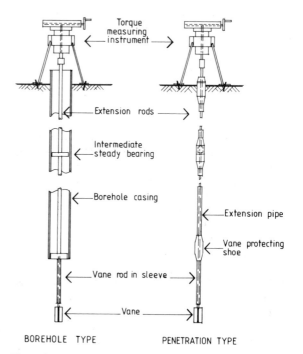

BOREHOLE TYPE   PENETRATION TYPE

Figure 2.14 Vane test apparatus

Table 2.5   Classification of fine-grained soils

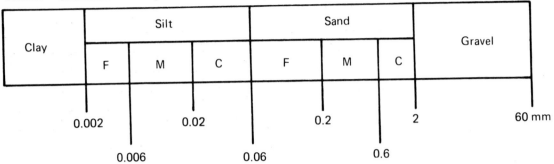

to cause a cylindrical surface of rupture and the diameter and height of the vane.

Grading test

The purpose of a grading test is to determine the particle size distribution of a soil. Soils, for the purpose of particle size classification, are grouped as follows:

(a) Fine-grained soils. Soils containing not less than 90% passing a 2.36 mm BS sieve.
(b) Medium-grained soils. Soils containing not less than 90% passing a 20 mm BS sieve.
(c) Coarse-grained soils. Soils containing not less than 90% passing a 40 mm BS sieve.

Group (a) are further classified as shown in Table 2.5.

The analysis for the above classification is carried out in two stages.

The separation of coarser fractions by sieving on a series of BS sieves.

The determination of the proportions of the finer particles by a sedimentation process, generally known as 'wet analysis'.

Sieving

The sample of soil is oven dried for 24 hours at 105–110°C. The sample is then weighed and sifted through a series of BS sieves (sizes 2.36, 1.18 mm, 600, 300, 212 150 and 75 μm); the amount of soil retained in each sieve is then weighed. To illustrate the recording and presentation of the results an example of a sieve analysis is given in Table 2.6.

A grading curve is then drawn by plotting the percentage of material passing each sieve against BS sieve sizes used. This is illustrated below in Figure 2.15.

Wet sieving

The sample is prepared as before by oven drying and weighed. The sample is then sifted through a 20 mm BS sieve. The remainder of the original samples that passes through the 10 mm sieve is then reduced to a convenient amount (for maximum sieve loads see Form 'F' BS1377 (1975)) through a riffle box. The weight of the sample is then recorded.

The sample is placed on a tray or in a bucket and covered with water. A solution of sodium hexametaphosphate is added at a rate of 2 g per one litre of water, the whole being well stirred. The sample is then left for one hour in the solution but stirred frequently.

After one hour the sample is washed through a 5 mm BS sieve into a tray, the fraction retained on the 5 mm sieve is to be kept for oven drying and weighing. The remainder of the sample is then washed through the

Table 2.6   Typical result of grading test

B.S. sieve size	Weight of material retained on sieve	Weight of material passing each sieve	% of material passing sieve
10 mm	6.47 g	168.13 g	96.3
5 mm	12.57 g	155.56 g	89.1
2.36 mm	19.73 g	135.83 g	77.8
1.18 mm	21.82 g	114.01 g	65.3
600 μm	35.61 g	78.40 g	44.9
300 μm	37.54 g	40.86 g	23.4
212 μm	9.44 g	31.42 g	18.0
150 μm	5.75 g	25.67 g	14.7
75 μm	9.96 g	15.71 g	9.0
< 75 μm	15.71 g		
Total wt.	Σ 174.60 g		

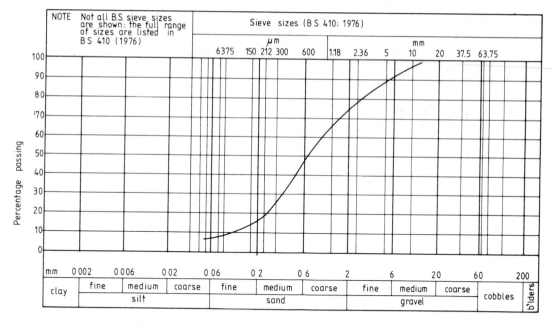

Figure 2.15 Graphical representation of grading

rest of the sieves made up of the following BS sieves:

2.36 mm, 1.18 mm and 75 μm

The fraction of the sample that passes the 75 μm sieve is then discarded and the process repeated until the water after washing through the sieves is virtually clear.

Finally, the remaining fractions of the sample are placed in separate trays and oven dried at 105–100°C for 24 hours. The dried samples are then sieved in the same manner as for the sieving test and the results recorded and presented as previously illustrated.

Sulphate content of soil and soil water

The presence of soluble sulphates in soils can cause the breakdown of Portland cement concretes in contact with water in such soils, due to the formation of calcium sulphurluminate. Due to an expansion of volume accompanying the formation of the compound, internal stresses are set up sufficient to disrupt the cement matrix and cause mechanical failure of the material as a whole.

Expressed in terms of sulphur trioxide the presence of less than 0.2% SO_3 in the soil or soil water is unlikely to cause any problems but should the content of SO_3 be 0.5% or above there is a serious risk of deterioration of the concrete.

Soluble sulphates are also a major factor in the corrosion of metal pipes laid in waterlogged clay soils.

The procedure and calculations for the determination of water soluble sulphate content are fully described in BS1377 (1975). Test 9 describes total sulphate content of soil and Test 10 describes the determination of sulphate content of groundwater and aqueous soil extracts. When obtaining results from either test it should be remembered that the moisture content of soils will fluctuate according to the time of the year, and hence the sulphate content.

Table 2.7 Approximate densities of soils

Material	Density kg/m³	Material	Density kg/m³
Chalk	2000	Gravel, with sand	1700
Clay, dry	1600	Loam, dry	1600
Clay, wet	2000	Loam, wet	2100
Earth, dry loose	1300	Mud	1600
Earth, dry rammed	1750	Pebbles	1750
Earth, damp	1600	Sand, dry	1500
Earth, very compact	1850	Sand, damp	1750
Earth, wet rammed	2000	Sand, wet	1900
Gravel	1750	Stones, broken	1600

Table 2.8   Maximum safe bearing capacities of soils

Material	MSBC kN/m²	Material	MSBC kN/m²
Limestone	1280	Clay, soft	55–110
Slate	3800	Clay, very soft	0– 55
Shale, hard	2140	Gravel, compact	430–640
Sandstone, soft	2140	Gravel, loose	215–430
Clay/shale	1070	Sand, compact	215–430
Boulder clay,	430–640	Sand, loose	110–215
Clay, stiff	215–430		
Clay, firm	110–215		

Identification of soils

Identification requires a good knowledge of soil mechanics which is beyond the scope of this book. However, there are certain ways of obtaining a general impression of the expected performance of soils. Tables 2.7, 2.8 and 2.9 will be of assistance.

Site report

So far we have been concerned with the practices of soil investigation. This will form only part of the information that will be required by the managing consultants to the contractor, before a start can be made on the construction of the intended structure. The complete report, that will also contain the soil investigation, is referred to as the 'site report'.

Table 2.9   General basis for field identification and classification of soils

		Size and nature of particles	
		Principal soil types	
	Types	Field identification	Composite types
Coarse-grained non-cohesive	Boulders Cobbles	Larger than 200 mm dia. Mostly between 200 and 75 mm	Boulder Gravels
	Gravels	Mostly between 75 mm and 2.36 mm BS sieve	Hoggin Sandy gravels
	Sands, uniform and graded	Composed of particles mostly between 2.36 mm and 75 µm BS sieves, and visible to the naked eye. Very little or no cohesion when dry. Sands may be classified as uniform or well graded according to the distribution of particle size. Uniform sands may be divided into coarse sands between 2.36 mm and 600 µm BS sieves, medium sands between 600 µm and 212 µm BS sieves and fine sands between 212 µm and 75 µm BS sieves.	Silty sands Micaceous sands Lateritic sands Clayey sands
Fine-grained cohesive	Silts, low plasticity	Particles mostly passing 75 µm BS sieve. Particles mostly invisible or barely visible to the naked eye. Some plasticity and exhibits marked dilatancy. Dries moderately quickly and can be dusted off the fingers. Dry lumps possess cohesion but can be powdered easily in the fingers.	Loams Clayey silts Organic silts
	Clays: Medium plasticity	Dry lumps can be broken but not powdered. They also disintegrate under water. Smooth touch and plastic, no dilatancy. Sticks to the fingers and dries clowly. Shrinks appreciably on drying, usually showing cracks.	Boulder clays Sandy clay Silty clays Maris Organic clays
	High plasticity	Lean and fat clays show those properties to a moderate and high degree respectively.	Lateritic clays
Organic	Peats	Fibrous organic material, usually brown or black in colour.	Sandy, silty or clayey peats

It is recommended that a comprehensive site report with information on the following items, if and when relevant, should be supplied.

General information:

(a) Ordnance and Geological maps; ✓
(b) General survey: i.e. location of site, corrections on existing OS maps, property lines etc.; ✓
(c) Approaches and access; ✓
(d) Restrictions: i.e. legal restrictions or existing structures or works beneath the site, rights of way etc.; ✓
(e) Drainage and sewage; ✓
(f) Supply of services: i.e. electricity, gas, water etc. ✓

Special information required for design and construction:

(a) Detailed survey;
(b) Present and past uses of the site;
(c) Soil investigation report;
(d) Hydrological report in the case of river and sea works;
(e) Climatic conditions;
(f) Sources of local materials for construction;
(g) Disposal of waste materials, i.e. distance to and availability of spoil tips.

Many contractors use pre-printed forms to be filled in as the information is acquired. This will help to avoid omissions which could prove very expensive at a later date. Part of a typical site investigation report form is shown in Figure 2.16.

Table 2.9 *(cont'd)*

	Strength and structural characteristics		
Strength		*Structure*	
Term	*Field test*	*Term*	*Field identification*
Loose Compact	Can be excavated with spade. 50 mm wooden peg can be easily driven. Requires pick for excavation. 50 mm wooden peg hard to drive more than a few centimetres.	Homogeneous	Deposits consisting essentially of one type.
Slightly cemented	Visual examination. Pick removes soil in lumps which can be abraded with thumb.	Stratified	Alternating layers of varying types.
Soft Firm	Easily moulded in the fingers. Can be moulded by strong pressure in the fingers.	Homogeneous Stratified	Deposits consisting essentially of one type. Alternating layers of varying types.
Very soft Soft Firm Stiff Hard	Exudes between fingers when squeezed in fist. Easily moulded in fingers Can be moulded by strong pressure in the fingers. Cannot be moulded in fingers. Brittle or very tough.	Fissured Intact Homogeneous Stratified Weathered	Breaks into polyhedral fragments along fissure planes. No fissures. Deposits consisting essentially of one type. Alternating layers of varying types. If layers are thin the soil may be described as laminated. Usually exhibits crumb or columnar structure.
Firm Spongy	Fibres compressed together. Very compressible and open structure.		

7 DEMOLITION OF EXISTING BUILDINGS	If any shoring and/or underpinning required to adjacent buildings and liability for their condition	

8	Details		Load Capacity	Width
ACCESS TO SITE	Existing roads on site			
	Existing roads adjacent to site			
	Temporary roads required			
	Access difficulties			

9	Strata expected to be found during excavation					
GROUND CONDITIONS. (STATE MAX. DEPTH AND LOCATION OF STARRED ITEMS).	Boreholes*	Boreholes Existing YES/NO			Boreholes Required	YES/NO
	Effect of Adverse Climatic Conditions on Site Operations					
	Water	Surface YES/NO	Depth of Table (m)	Tidal YES/NO	Springs* Yes/No (If yes give number)	Pumping required YES/NO
	Any existing foundations, roots, rock or any other material likely to be found in ground					
	Stability of excavations					
	Timbering required					
	Any other ground details*					

Figure 2.16   Extract from site investigation report form

Site layout

If a contract is to be carried out in an efficient manner, an essential element of this goal is the planning of the site layout, before the contract is started. No two sites are the same and 'perfect' solutions do not exist, hence the site plan adopted for a contract will be an optimum solution to a problem that has many variables to be taken into account. The aim of a site planning exercise is to produce a layout that is logical, orderly and above all practical.

In general terms, the main items that must be considered in laying out a new site may be listed as follows:

(i)   Access to site and 'on site' roads;
(ii) Storage of materials;
(iii) Plant requirements and movement of plant;
(iv) Temporary buildings for the contractors' team and other persons resident on the site;
(v)   Temporary services;
(vi) Fencing and hoardings.

(i) Access to site and 'on site' roads

Access to the site would normally be described in the contract. If the site has more than one point of access each entrance should be identified by either a letter or number and instructions clearly displayed as to where visitors and material deliveries should report on arrival. To facilitate on-off movement of site traffic, a system of exit and entrance only gates can be of advantage.

The movement of plant on a site must be planned for the efficient and economic operation of the machines. This is particularly so when siting haulage roads to spoil tips. The decision on what type of roads are to be used, e.g. rough access, waterbound or bitumen sprayed, will depend upon the type of plant that is to be used, the ground conditions of the site and the cost effectiveness.

For example, if the first two points are considered, tracked vehicles fitted with grips cannot pass continuously over waterbound or bitumen sprayed roads, as the grips on the tracks will destroy the surface. Heavy plant on flat tracks, such as diggers, cannot pass over soft or hard ground without seriously damaging the surface. Dump trucks apply high concentrated wheel loads and, therefore, cannot pass over soft ground and so on.

Generally rough access roads are used by on-site traffic and waterbound or bitumen sprayed roads for on-off site traffic. These roads should be maintained to prevent the breaking up of the haunches and ditches and potholes occurring that could damage wheeled vehicles.

The layout of site roads should provide for a smooth movement of all site traffic with an economy of distance. Where off site vehicles cross over or use on site vehicle roads, warning signs must be posted or be supervised for reasons of safety.

(ii) Storage of materials

For normal site conditions, the materials to be used for concrete production are delivered to the site ready for batching and mixing. It is essential that the materials are not allowed to deteriorate after being delivered to the site.

Cement is usually delivered to the site in 50 kg bags. This is a convenient size for handling and storage, although increasing use is being made of bulk handling equipment. The advantage to be gained from this method of storage is dependent upon the size of the site and the amount of concrete to be produced.

When cement is delivered to site in bags it should not be stored directly on the ground, as it would soon become damp. The bags should be placed on a raised platform and covered with waterproof sheets. The deliveries should be so arranged that the cement is not subjected to prolonged storage.

Aggregates are delivered to site in at least two sizes. These are classified as coarse and fine. It may be necessary for the coarse aggregate to be delivered in a larger number of different sizes. These could be 40 mm, 20 mm and 10 mm depending upon the use required on site. The aggregates should be kept separated on the site by providing division walls, each size being clearly labelled. The ground on which the aggregates are to be stored should be given a layer of lean concrete, laid to falls. This is to ensure that the aggregates do not become contaminated with mud.

All materials should be stored in an orderly fashion. A great deal of wastage can occur if the storage methods are lax. As a general rule, the storage of materials should be so positioned as to avoid double handling and awkward access for delivery trucks.

(iii) Plant requirements and movement of plant

For the choice and use of excavating plant, see Chapter 3.

In general terms the location and working procedure of plant should aim to minimise the re-positioning of the machine, and maximise the coverage of the area that the plant is required to operate over. This is particularly so for cranes.

Tower cranes, unless mounted on rails, derricks and cableways, should not have to be re-positioned during a contract as the process is costly and, of course, during re-positioning work, the crane will be out of action. Excavating plant, particularly on extended sites such as roads and airfields, should 'work as they go' – for example, a scraper should start its run directed towards the spoil tip. Concreting plant should be positioned to give minimum delivery distance to the areas of the site that require the bulk of the concrete output.

(iv) Temporary buildings

The contractor's main office and the general storage buildings for site equipment should be located near the main entrance to the site. As all materials being delivered to the site must be checked on arrival, the storekeeper's office must be at the main entrance. Care must be taken to avoid obstructing the entrance to other vehicles while checking is carried out. For good communications it is wise to establish the resident engineer's office close to the contractor's office. If room allows, the sub-contractor's office could also be in this location. This will lead to an economy in providing such services as telephones, heating, lighting, cleaning etc.

By following these points an administrative area will be logically situated around the main entrance to the site thereby decreasing the number of persons walking or driving around the construction areas. A final point is that where feasible, the contractor's and resident engineer's offices should be positioned to allow an overall view of the site from their windows including the main entrance point.

(v) Temporary services

It is the contractor's duty to arrange the supply of all temporary services such as telephones, electricity and sanitation. Consideration must also be given to the diversion, due to the contract, of any existing services. It is usual for the engineer to supply the contractor with sufficient detailed information to locate the services in question, and contact any interested bodies such as British Telecom, Post Office, local water and electricity authorities etc. It will also be necessary to arrange who will be responsible for the diverting of any services, i.e. the contractor or authority concerned.

(vi) Fencing and hoardings

Fencing and hoardings serve two purposes:

(a) to control the delivery and removal of materials from the site by controlled entry and exit points, and

Figure 2.17   Dragline

(b) to prevent unauthorised persons entering the site. In rural areas this would also apply to livestock.

It is the contractor's duty to erect and maintain the boundary fencing or hoardings. The type chosen will depend upon the contractor's preference and the location of the site. On certain sites, particularly in cities, extra security may be required. In such cases, professional security firms may be hired by the contractor.

Concrete production plant

The development of plant for concrete production has made considerable progress in recent years and this trend is in line with the general mechanisation of the construction industry.

The following factors should be taken into account when selecting plant for concrete production:

(i)   The total volume of concrete required;
(ii) Amount of space available for setting up plant;

(iii) Quality of concrete required;
(iv) Site conditions (restrictions, noise, soil type etc.).

Types of plant include:

Handling equipment

Bulk cement handling requires the use of a completely enclosed conveying system on the site to move cement from the delivery vehicles to the storage silos or bins. Some bulk cement vehicles carry the cement in pressurised tanks. This system is designed to transfer the cement from the delivery vehicles to storage bins which may be up to 50 m from the delivery point. Cement may also be elevated into silos by this method.

Aggregates may be conveyed to the batching and mixing plant either by cranes, draglines or high lift shovels – see Figure 2.17. Conveyor belts or a form of bucket elevator can also be used.

Draglines are commonly used for large jobs to transfer aggregates from stockpiles to the overhead hoppers of the batching plant. A manually operated dragline

Figure 2.18   Site batching plant

shovel which is attached to the mixer is used on some sites to pick up aggregates and transfer to the mixer skip.

Batching plant

It is usual to use a separate weighing mechanism for the cement as this practice improves the accuracy of batching operations. Portable cement silos are widely used and these incorporate a separate weighing device for the cement – see Figure 2.18. Several types of batchers are used for the aggregate, one such type is the swinging weigh batcher which consists of two weighing hoppers which rotate about a central vertical axis. In this way two batches of material can be weighed simultaneously. It should be noted that the weighing mechanism will require frequent checks to ensure the accuracy of the aggregate batching.

Mixing plant

There are four main types of batch mixers – non-tilting drum, tilting drum, reversing drum and paddle mixers. BS1305 designates mixers by a number representing the nominal batch capacity in litres and a letter indicating the type of mixer.

The commonest size of mixers are 100, 150, 200, 250, 350, 500 and 750 litres and can be designated as follows:

100NT, 150T, 200P etc.,

where
  NT = non-tilting
   T = tilting
   P = paddle

Figure 2.19   Powered dumper

It is usual for the mixer to have a maximum capacity of 10% more than its normal capacity.

The mixing time varies from about 80 seconds to about 2 minutes. Pan or paddle mixers give very consistent mixes and are used for high grade concrete work. Most mixers, except those of the smallest type, have a water tank incorporated which measures the water for each batch.

Transporting concrete

Any plant used for moving freshly mixed concrete from the mixer to the point of placing must ensure that this operation is done as quickly as possible and that the concrete does not dry out, segregate, or become compacted before it can be placed (30 minutes from mixing to placing).

Types of plant consists of:

(i) Wheelbarrows
(ii) Dumpers
(iii) Mono-rail transporters
(iv) Hoists
(v) Crane skips and buckets
(vi) Chutes
(vii) Conveyor belts
(viii) Pumping

(i) Wheelbarrows

These should be of such a capacity that they are able to accommodate a complete batch of concrete; although barrows are usually too small to allow a single complete discharge of concrete, but they are convenient to use for carrying concrete along narrow boards.

(ii) Dumpers

The powered dumper is a variation of the wheelbarrow theme – see Figure 2.19. Since they have two pairs of wheels and a lower centre of gravity, the dumper is more stable. It can carry more material than the wheelbarrow and obviously the transportation and placement of the concrete is quicker.

(iii) Mono-rail transporters

These are very convenient for use with small and medium-sized mixing plants. They have the advantage that they can carry concrete across difficult sites without requiring much space. The rail is easily fixed and can be moved to a new position as the work proceeds – see Figure 2.20.

Figure 2.20 Loading a monorail conveyor from a dumper truck

(iv) Hoists

Hoists are becoming more general on construction sites and are available with skips which carry the concrete vertically, discharging into receiving hoppers at each working level which, in turn, feed wheelbarrows. Hoists are capable of transporting 13 m^3 of concrete per hour to a height of 150 m.

(v) Crane skips and buckets

When a crane is used for moving concrete, a suitable skip or bucket is required. The mix can be delivered directly to the point of placing in a single operation without intermediate handling – see Figure 2.21. Large and small quantities can be handled to almost any height. This method is, therefore, the most efficient means of handling concrete for most structural work.

(vi) Chutes

These are used generally in conjunction with ready mixed concrete deliveries. The advantage being that deliveries can be made adjacent to the point of placing and the chute will transfer the concrete. This is particularly useful in the case of concreting foundations – see Figure 2.22.

(vii) Conveyor belts

These are available in a convenient form which allows them to be easily moved to different parts of the site. When they are joined together a continuous flow of

Figure 2.21   Placing concrete using a skip

Figure 2.22   Ready-mixed concrete with a discharge chute

Figure 2.23   Concrete pumped by 32 m boom from truck mixer

Figure 2.24 Concrete truck mixer

concrete can be made from the mixer to the point of placing. The concrete must be fed onto the belt at a uniform rate otherwise there is a tendency for the belt to stop due to overloading.

(viii) Pumping

This is a very successful method of transporting concrete across congested sites where the mixing plant cannot be positioned close to the point of placing – see Figure 2.23.

Pumping concrete is more economical when reasonably large quantities are required and special attention should be paid to the concrete mix to ensure a smooth flow of concrete.

The speed of pumping exceeds most other methods of placing by a significant margin, which is the only justification for using this expensive method. An alternative method is the use of a pneumatic placer which forces the concrete through a pipeline by compressed air. Most of the problems associated with pumping concrete apply to this method.

Ready-mixed concrete

It is often more economic to bring concrete to the site already batched and mixed, particularly if site conditions are very congested or where the concrete is required in different places as the work proceeds, as, for example, in road construction or large foundations.

Ready-mixed concrete costs more per cubic metre than site-mixed, but savings in labour and prevention of waste reduces the cost.

Ready-mixed concrete can be delivered to a site up to 10 miles or about 30 minutes from the depot. Producers of ready-mixed concrete use a truck mixer for delivery – see Figure 2.24. The batched materials are added dry at the depot, and the water is added during the journey, just before reaching the site.

Site lifting or hoisting plant

In addition to transporting concrete horizontally as previously discussed, it is often necessary to lift or hoist concrete and other materials to a working area well above ground level. The type of plant used for these operations are generally cranes and hoists.

Cranes

This type of plant may be divided into two broad groups: mobile and stationary cranes. When selecting the type of crane required for a contract, consideration should be given to the following:

(i)   The length of time for which the crane will be required.

(ii)   The weights and dimensions of loads, the distances and heights they have to be moved.

(iii) The frequency of lifts and the distances between the positions from which loads have to be picked up.

(iv)   Special requirements such as heavy loads, rapid lowering, or exceptionally high lifts.

(v)   Details of ground conditions and the nature of the work being carried out in the area in which the crane will operate, particularly if mobiles are used.

(vi)   The amount of space available for the erection and dismantling operations if a tower crane is being considered.

Mobile cranes

This type of crane is available in a wide variety of designs and capabilities, generally with a 360° rotation or slewing circle and capable of being moved to different locations on the site or between sites.

Mobile cranes can be divided into four groups as follows:

(i)   Mobile wheel cranes
(ii)   Lorry or truck-mounted cranes
(iii) Track-mounted cranes
(iv)   Gantry cranes

(i) Mobile wheel cranes

The most simple type is a crane mounted on motorised wheels – see Figure 2.25. The lifting capacity is generally low, with a maximum lift of about 10 tonnes. They are ideal for operating in a precast concrete casting yard or similar area where there is a hard level surface.

(ii) Lorry or truck-mounted cranes

A further development is the crane mounted on a specially designed lorry or truck. The advantage of this type is the fact that they can travel easily between sites thus making them fully mobile, but to operate efficiently they require a firm and level surface.

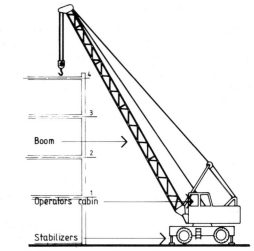

Figure 2.25   Mobile wheeled crane

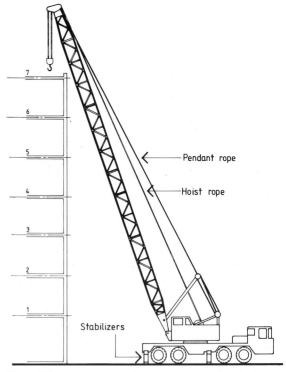

Figure 2.26   Lorry-mounted crane

The capacity of these cranes is about 20 tonnes maximum, the reach may be increased by the use of outriggers – see Figure 2.26.

(iii) Track-mounted cranes

This type consists generally of excavators which have been rigged as cranes. The general purpose types have a maximum capacity of about 10 tonnes, but a machine which has been made exclusively for crane duties has a maximum lift of about 45 tonnes – see Figure 2.27.

The cranes can traverse most sites without undue difficulty, but there is a need for low-loading transport to move them between sites.

(iv) Gantry Cranes

Two sets of rails are installed, one each side of the working area, and the gantry or portal crane can then transverse the full length of the structure.

The lifting gear is suspended from the horizontal frame and is capable of moving the full width of the portal.

Although the use is somewhat limited, it is particularly useful to assist in the construction of a medium rise structure of considerable length, thereby making full use of the transversing rails.

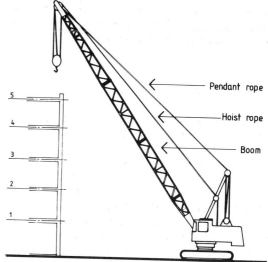

Figure 2.27   Track-mounted crane

Stationary cranes

These can be defined as cranes of various types which are firmly fixed to some form of base at their working positions and also cranes which have limited lateral movement restricted to the immediate vicinity of their working positions.

Stationary cranes are therefore classified as derrick cranes and tower cranes.

Derrick cranes

These are cranes fixed at their working position and used mainly on large civil engineering projects because of their lifting capacity and wide reach.

There are two types of derrick crane:

(a) Scotch derrick

This consists of a slewing mast and a luffing jib. Stability is provided by using lattice steel backstays which run from the top to the extremes of two base sleepers. The stays limit the slew of the crane to about 270°. This type of derrick can also be mounted on rails thus increasing the flexibility of the crane.

The scotch derrick is ideal for lifting heavy loads in 3–20 tonne range.

(b) Guyed derrick

This is a simple form of stationary crane, consisting of a vertical guyed mast with a pedestal bearing stabilised by a number of anchored guy ropes. As the jib is slightly shorter than the mast, it can rotate through

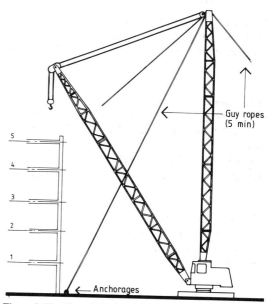

Figure 2.28   Guyed derrick

360°. They are quickly and easily erected and dismantled and removed from one position to another – see Figure 2.28.

This type of crane can be used for erecting large areas of structural steelwork.

Tower cranes

Power driven tower cranes are available in a variety of types and configurations, according to the type of tower, jib and base employed.

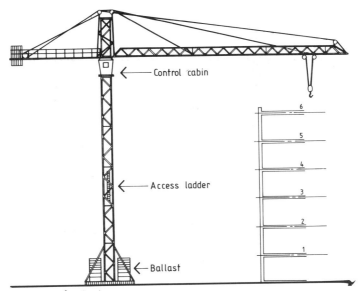

Figure 2.29   Static tower crane

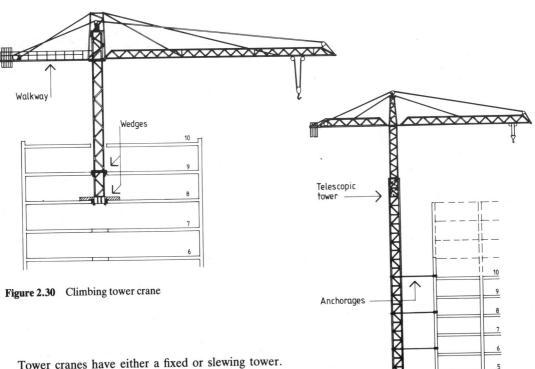

Figure 2.30   Climbing tower crane

Tower cranes have either a fixed or slewing tower. With the former, the slewing ring is situated at the bottom of the tower and the whole of the tower and jib slews relative to the crane base. Tower cranes can be divided into four principal types:

(i)   Static tower cranes
(ii) Rail-mounted tower cranes
(iii) Truck-mounted tower cranes
(iv) Crawler-mounted tower cranes

Figure 2.31   Supported static tower crane

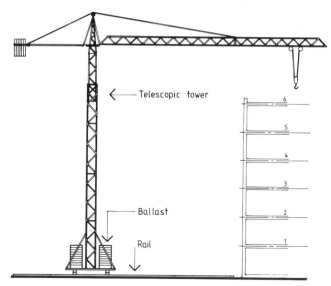

Figure 2.32   Travelling tower crane

(i) Static tower cranes

There are several types of static tower cranes. They include the self-supporting tower crane which is mounted on its own static base which is constructed of insitu concrete, or a special tower base cast into a foundation. They generally have a greater lifting capacity than other types of crane – see Figure 2.29.

They should be positioned in front or to one side of the structure under construction and are preferred when the site is very congested. Another example is the climbing crane, which is similar in construction to the self-supporting type, but is used for lifting to much greater heights – see Figure 2.30.

The crane is fixed to the structure during construction by means of climbing frames and wedges. The height of the crane can be extended as the structure increases by means of climbing ladders attached to the frames. The crane is usually mounted initially on its own fixed base and its support later transferred to the climbing frames and ladders.

A further example is the supported static tower crane which is sometimes used if the loads to be lifted are in excess of that possible by self-supporting tower cranes. As the tower is tied to the structure, see Figure 2.31, undue stresses may be set up in the structure. These should be allowed for at the design stage.

(ii) Rail-mounted tower cranes

This form of crane is supported on heavy wheeled bogies which in turn are mounted on a wide gauge rail track. The track is laid on conventional sleepers supported by well graded ballast. The level of the track

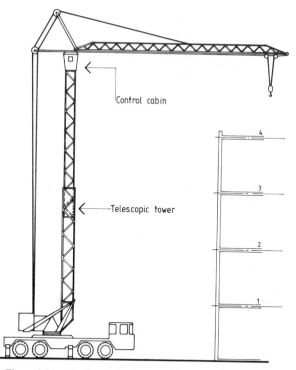

Figure 2.33   Truck-mounted tower crane

must be checked frequently to ensure the crane remains stable. This type of crane may have a fixed or slewing tower and various types of jib – see Figure 2.32.

(iii) Truck-mounted tower cranes

Sometimes it is necessary to use a tower crane mounted on a specially adapted lorry or truck. This type is ideal for use on low rise construction when a static tower crane cannot be justified. It is essential that the outriggers are extended and in position before commencing lifting operations. To facilitate ease of transport, this type of crane is fitted with telescopic towers – see Figure 2.33. Small capacity tower cranes are available on wheel mountings. These are not usually self-propelling, but moved by towing behind a suitable vehicle.

This type is fitted with stabilisers or outriggers and jacks which should be set before commencing lifting operations. It is essential that the wheels are lifted clear of the supporting surface when operating.

(iv) Crawler-mounted tower cranes

There are two principal types of crawler base used on this type of machine. One is the twin track type, which is mounted on one pair of crawler tracks and requires extended outriggers and jacks to retain stability when lifting. The other is the straddle type which is mounted on four widely spaced crawler tracks each of which can be adjusted for height.

Both types should be set firm and level when handling loads. They usually have a telescopic tower but a variety of jibs may be used.

Crawler-mounted tower cranes are required to be transported on a low-loader or towed on special road axles when travelling on public highways.

Another type of crane in common use is a combination of the derrick and tower crane known as a *monotower crane*. This is basically an elevated derrick crane supported by a well braced tower. To ensure stability the mast of the crane extends well down the tower to a pivot bearing. The crane is slewed from a ring mounted on the top of the tower through a full 360° and for construction purposes the capacity is 5 to 20 tonnes with jibs up to 40 m long.

This type of crane is limited to large civil engineering contracts where the high cost of erecting the crane together with the expensive foundation required can be fully justified.

Erection and dismantling of cranes

All operations for erecting and dismantling cranes are set out in BS 3010 (1972).

Also, particular attention should be paid to the manufacturers' instructions on the erection and dismantling procedures for cranes.

Hoists

This type of plant is used to transport materials or passengers vertically by means of a moving level platform operating between vertical guides. The guides are built in sections so they may be extended as the structure rises. They are usually tied back to the scaffolding to provide stability. The hoists are manufactured in sizes up to 1.5 tonnes capacity and although their main use is to transport general construction materials, the platform can be modified to accept a skip to move concrete as mentioned previously. In some high rise constructions, hoists are used to transport passengers to high working areas.

On low rise construction, advantage can be made of a mobile hoist with a maximum height of 24 m. In this case, the guides are fitted to a central vertical lattice mast which is mounted on a wheeled chassis. This type does not require tying to the structure unless extension pieces are added.

The Construction (Lifting Operations) Regulations 1961

Regulations No. 10 to No. 41 inclusive of Parts III and IV of the above regulations, set out in detail the statutory crane requirements with reference to the following: (a) lifting gear; (b) erection, fixing and anchoring, lifting appliances; (c) travelling and slewing of cranes; (d) platform sizes and cabins; (e) drums, pulleys and brake controls; (f) examination of equipment; (g) stability; (h) operating requirements; (i) erection and dismantling of cranes; (j) training of operators; (k) testing and inspection of equipment; (l) safety.

In the case of hoists, Regulations No. 42 to No. 49 inclusive of parts V and VI set out in detail the statutory hoist requirements with reference to the following: (a) enclosure of hoists; (b) operating and controlling hoists; (c) safety of hoists in operation; (d) testing; (e) lifting devices for passengers; (f) safety of passengers; (g) security of loads.

3 Earthworks

Groundwater control

In excavation work, the ingress of surface and ground water can cause serious problems such as the under-mining of supports to the sides of the excavation and the very costly delay of the work being held up due to flooding. There is also the general problem of working in waterlogged conditions; it is, therefore, important to control, or exclude, the flow of surface and groundwater in the excavation.

The methods used to control groundwater are generally termed as geotechnical processes, of which there are many. Such methods include the lowering of the groundwater, changing the physical characteristics of the sub-soil, freezing the groundwater or the use of compressed air.

Before a final choice can be made on the techniques to be used, a good knowledge of the ground and water conditions of the site, and possibly surrounding area, will have to be obtained. This must be carried out before work is commenced, as unnecessary delays will be caused if the problem of groundwater is only considered when difficulties arise during the construction work.

Preliminary site investigation

As has been previously described, a site investigation is carried out to obtain a general picture of the underlying subsoil strata and water level of the site. The investigation will establish the likelihood of the need for dewatering operations; if this is seen to be so, then a more detailed investigation will be required.

Detailed investigation

This investigation is required to allow a detailed study of the soil profiles, groundwater conditions and, if relevant, a general history of other excavation and foundation works that have been carried out in the area. This latter point can be of great use in assessing the problems to be faced.

The types of soil that make up the underlying strata and their particle size distribution will have an important bearing on the choice of dewatering method to be used. When fine grain soils are encountered it is recommended to obtain continuous samples.

Particle size analysis should be made on representative samples from all silts, sand and gravel layers. In the case of silts and fine sands the samples should be undisturbed.

If dewatering is to be carried out in the vicinity of existing structures, the compressibility of the various strata must be determined in order to ascertain whether excessive settlement is likely to occur. The settlement is caused by the reduction of pore water pressure in a compressible strata. If this were to happen, serious damage may be caused to the existing structures due to the settlement of their foundations.

A study of the groundwater flow pattern is of great value when considering the layout of the dewatering installations. The purpose of the study is to establish the source of the groundwater and how it flows in the region of the site. Groundwater is normally found in pervious strata due to the percolation of rainfall and run-offs and near streams and rivers.

There are, for the purposes of civil engineering, two types of groundwater problems:

(a) Confined
(b) Unconfined

(a) Confined

Groundwater is said to be confined when the layers in which it lies has along its upper flow boundary a layer of low permeability. If the source of the groundwater is above the level of the area of investigation, the water head in the pervious layer at this point may be above the ground level. This condition is known as artesian, see Figure 3.1.

(b) Unconfined

When the upper flow boundary of the groundwater is not confined by a layer of low permeability it is said to be unconfined. The ground water pressure at this boundary will be equal to atmospheric pressure, see Figure 3.1.

In the case of confined groundwater it is necessary to measure the water pressure in the pervious strata at suitable intervals across the site in order that the maximum water pressures be ascertained. It is also required to build up a picture of the groundwater flow across the site: to act as a control to measure changes in water pressure due to pumping tests, and groundwater lowering operations. Also it ensures that the groundwater pressure does not exceed values which could cause damage to temporary or permanent works.

Groundwater pressure is measured by devices known

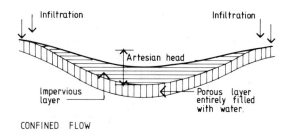

CONFINED FLOW

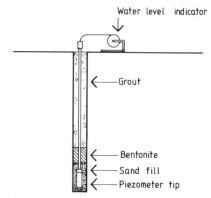

Figure 3.1   Confined and unconfined flow

as piezometers. The apparatus basically consists of a ceramic or porous plastic piezometer tip which is connected to a PVC or steel tube standpipe. The water pressure is measured by either connecting the standpipe to a water level indicator or a monometer gauge, a device for measuring water pressure difference.

If the apparatus is to remain insitu during the construction work and a borehole was used to position it, the area of the tip should be filled with sand and sealed with a layer of bentonite. The borehole is then filled in or grouted up, see Figure 3.2. It should be noted though the choice of piezometer tip does require a specialist knowledge.

It is also necessary to measure the rate of flow and permeability of the groundwater. The rate of flow is controlled by the permeability of the soil and is expressed by the equation (Darcy's Law):

$$v = ki$$

where:

v = the superficial velocity of flow through the soil
i = the hydraulic gradient
k = the permeability

k is measured in m/s and depends chiefly on particle size and grading. Typical values of permeability for soils range from 1×10^{-5} m/s for coarse sands to 1×10^{-9} m/s for a clay.

Permeability can be measured either insitu or in a laboratory. The usual site tests are full scale pumping tests, rising and falling head tests in a borehole, and the calculation of flow conditions measured in boreholes.

Choice of groundwater control or treatment
Having obtained the above information, the final factors to be considered before a decision on the method to be adopted will include the size and shape of the excavation, the length of time it will be open and the overall economics of any particular choice. In arriving at a final solution the following possibilities should be considered.

Avoidance of groundwater
Instead of trying to control the ingress of groundwater into foundation excavations or, in some cases, permanent works, it may be advantageous to avoid working in waterlogged ground. For example, by using a shallow raft foundation above groundwater level or by transferring the loads to an underlying strata by the use of piles.

In some cases by rearranging the location of individual structures so that the deeper foundations are made in more favourable ground could be a better solution.

Exclusion of groundwater
The process of exclusion can be broadly divided into two main groups. **Temporary** – in this case the water table is lowered so that work can be carried out in excavations or other temporary situations; and **permanent** – when the water can be excluded from the working area by surrounding it by an almost waterproof barrier.

Figure 3.2   Diagrammatic view of piezometer insitu (casagrande method)

Methods of temporarily excluding groundwater

Pumping from sumps

Traditionally sumps are sited within the area of excavation, but when large quantities of water have to be pumped it may be more convenient to form the sump just outside the excavation area. This will also avoid the risk of damage to supporting timbers in the excavation by erosion of the soil at formation level.

Except on very small sites, at least two sumps should be used. The sump should be excavated to the full depth required to drain the excavation before the main excavation reaches the level of groundwater and maintains in its original form until completion of the construction work.

To prevent loss of soil due to pumping from the sump a filter medium can be placed in the sump. To do this a cage of perforated metal is placed in the bottom of the sump and the space between the cage and the ground filled with a graded gravel filter material. As the gravel is placed the timber or steel supports to the sides of the supports are withdrawn.

By excavating the sump before the main excavation below the groundwater level is commenced, the groundwater can be kept below excavation level at all stages and any difficulties that might arise due to the soil or groundwater will be identified during the construction of the sump. This will allow the engineer and contractor to make any changes in the excavation scheme before the main work begins.

Drainage channels should be dug to falls leading to the sumps to prevent water standing on the surface of the excavation. The falls must not be too slow to allow silting up or too steep to cause erosion. The most efficient method is to use open-jointed earthenware pipes in the channels backfilled with a filter material.

If the excavation is to be taken below a pervious layer into an impervious one it is good practice to have the sumps at the base of the pervious layer, being fed by channels stepped back into the perimeter of the excavations. This is known as a 'garland drain'. The advantage of doing this is that the formation level of the main excavation will be kept free of water which could lead to it being cut up by the passage of the plant, and a reduction in the size of pumps required – see Figure 3.3.

Land drains

Land drains are used on construction sites to control surface water over the site and, in certain cases, to prevent surface water run-off entering an excavation.

They are normally open joined porous clay, concrete or perforated pitchfibre pipes. Vitrified or polythene pipes are also used. The choice of pipe is mainly based

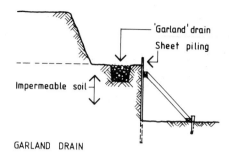

GARLAND DRAIN

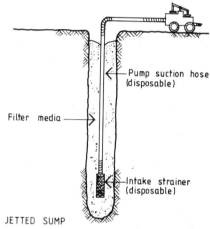

JETTED SUMP

Figure 3.3 Drain and sump

on the availability of any particular type. How they are laid in a trench is of importance though. Figure 3.4 shows a cross section through a typical land drain trench.

If porous pipes are used there is the risk of the pipes being silted up but with correct backfilling the risk is small. Figure 3.5 shows some typical layouts of land drain systems. The whole system will generally discharge into either a soakaway or a watercourse via a catchpit.

Clinker or rubble backfill to allow free drainage

Layer of inverted turf or straw to prevent particles silting up pipe if porous pipe used.

75 dia branch pipes and 100 dia main pipes.

figure 3.4 Typical section through land drain trench

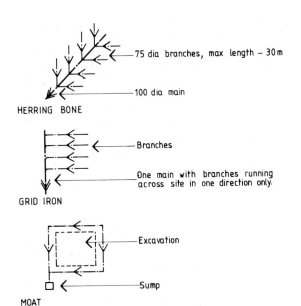

HERRING BONE

75 dia branches, max length – 30m

100 dia main

GRID IRON

Branches

One main with branches running across site in one direction only.

MOAT

Excavation

Sump

Figure 3.5   Layout of land drains

Pumping from wellpoints

A wellpoint is a mechanical device used as a small well that can be readily installed in the ground and withdrawn. The wellpoint system consists of a number of wells installed around an excavation. The wells are connected at the top to a header main which, in turn, is connected to a pump. The well point system has the advantage that water is taken away from the excavation by the action of the pump, thereby stabilizing the sides of the excavation.

The system can be quickly and cheaply installed and is particularly suitable for trench work. A disadvantage of the system is the limited suction lift. A lowering of about 5 m below pump level is considered as a practical limit, and for deeper excavations the wellpoints should be installed in two or more stages.

The spacing of wellpoints can vary from 0.30 m to 1.50 m depending on the permeability of the soil and can be installed in the ring or progressive system. The total capacity of a single pump is between 50 and 60 wells.

The ring system consists of wellpoints installed around the perimeter of an excavation and connected to a header main and pump, see Figure 3.6; this system is ideal for large rectangular excavations such as basement areas.

The progressive system is used for trench excavations with the header main laid along the side of the trench and pumping is continuously in progress in one length.

An alternative to the above is the horizontal method which is suitable for dewatering long trenches and can be laid to a depth of 6 m.

As mentioned previously, the limit for practical lowering of water is about 5 m below pump level. If deeper excavation below standing water level is required, a second or successive stage of wellpoints is required – see Figure 3.6.

Pumping from bored wells

Wells usually of about 300 mm diameter or greater are bored to a depth of about 10 m below pump level. Another tube is placed in the borehole which is provided with a perforated screen over the depth of soil requiring dewatering. The lower end of the inner tube is unperforated and acts as a sump for the fine material. The space between the two linings is filled with a suitable filler material over the length of the perforated section. The remainder of the space can be filled with any suitable material. Before the submersible pump is inserted in the tube the water is 'surged' to encourage the flow of water through the filter and wash out unwanted fine materials, see Figure 3.7.

The cost of installation of bored wells is relatively high, therefore the process is restricted to projects which have a long construction period.

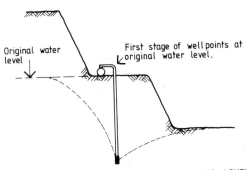

Original water level

First stage of wellpoints at original water level.

EXCAVATION OF BASEMENT TO NATURAL WATER LEVEL BEFORE POSITIONING WELLPOINT.

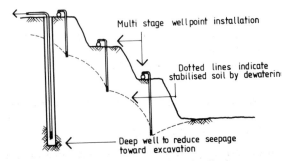

Multi stage wellpoint installation

Dotted lines indicate stabilised soil by dewaterin

Deep well to reduce seepage toward excavation

Figure 3.6   Combination of deep wells and wellpoint system

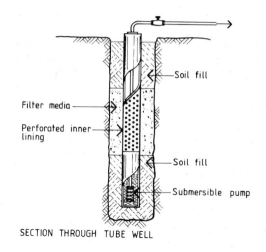

Soil fill

Filter media

Perforated inner lining

Soil fill

Submersible pump

SECTION THROUGH TUBE WELL

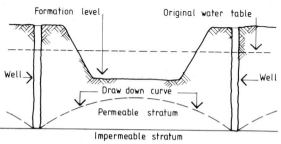

Formation level

Original water table

Well

Well

Draw down curve

Permeable stratum

Impermeable stratum

Figure 3.7 Wells used to lower water table to a level below formation of excavation

Electro-osmosis

Dewatering of silts and clays can present some difficulty if the traditional methods are employed. The principle of electro-osmosis is that soil particles carry a negative charge which attracts the positively charged ends of the water molecules creating a balanced stage. If this balance is disturbed the water will flow. The flow is achieved by inserting two electrodes into the saturated soil and passing a direct electrical charge between them. When the current passes through the ground between the electrodes it disturbs the balance and the water flows towards the negative electrode which is usually a wellpoint and the water is pumped away in the normal way.

The main disadvantage of this method is the high installation and running costs and, therefore, its use is limited to one of remedying a difficult situation.

Methods of permanently excluding groundwater

Cast insitu diaphragm walls

Water can be prevented from entering excavations by surrounding them with an impermeable wall construc-

ted insitu. The walls can vary from 450 mm to 1 m in width with a depth of up to 45 m.

The trench for the wall panels is generally excavated using mechanical grabs mounted on crawler crane rigs. As the trench is excavated, bentonite slurry is pumped in to ensure the trench remains stable. On completion of the excavated trench it is necessary to de-sand the bentonite slurry to ensure that soil contaminated slurry is not trapped in the wall reinforcement during concreting. The concrete is placed using a tremie pipe assembly complete with gate and hopper. As the concrete is poured the bentonite slurry rises in the trench and is progressively removed by pump for disposal or re-use.

A classic case of the use of diaphragm walls is the Holmesdale Tunnel situated on the new M25 London Orbital Road at Waltham Cross. This tunnel is 70 m long and was constructed using the cut and cover method. The walls for the tunnel are, on average, 14.7 m in depth and heavily reinforced. Excavation for the underpass took place after the installation of the wall panels, thus avoiding the necessity of using temporary supports to the sides of the excavation. Figures 3.8 and 3.9 show details of the construction work.

Sheet piling

This method is used to form a barrier to the ingress of groundwater. Sheet piling can be designed as a retaining wall (see section on retaining walls) or as a temporary enclosure for excavation works.

As this method involves the use of a drop hammer, the resulting vibration and noise may render this method as unsuitable especially if installation takes place close to existing structures.

Contiguous piling

As an alternative to diaphragm walls, concrete bored piles can be installed in a line. Care must be taken to ensure that the piles touch over their total length. To ensure water exclusion from an excavation, it may be necessary to inject grout between the piles. This method is used as a permanent retaining or basement wall and is completed by the addition of a reinforced concrete inner wall, terminated with a capping beam.

Grout injection

This method of water exclusion involves the injection of fine suspensions or fluids into the pore spaces, fissures or cavities in the soil or rock thus reducing the permeability. Grouting is a fairly costly process and it is, therefore, of paramount importance that the volume of the basic material is kept to a minimum. In this respect, chemical grouts have certain advantages. Examples and types of grout currently in use for treating various ground conditions are shown in Table

Figure 3.8   Construction work showing the diaphragm walls completed

Figure 3.9   Excavating for diaphragm wall

3.1. Grouting is usually undertaken by specialist contractors, but to generalise, grouting by injection is achieved by pumping a prepared grout to the head of steel or plastic tubes placed in the ground that has to be treated. It is essential that provision be made for control of the grouting operations by the engineer in charge of the works.

The freezing process

Because of its high cost, freezing the ground to prevent the ingress of water into excavations is generally used only when all other methods are impractical for one reason or another. Basically, the method involves the sinking of boreholes around the excavation at 1.0 to 1.5 m centres, the boreholes are lined with an inner or outer tube. The inner tube is open at the bottom and connected to the flow pipe at the top. The outer tube is sealed at the bottom and connected by a gland to the return pipe at the top. Chilled brine from a refrigeration plant is pumped through the tubes and eventually an ice wall is formed around the borehole. The method is slow, taking in some cases up to four months to freeze the ground satisfactorily. Considerable saving in freezing time can be achieved by the use of nitrogen fed directly from insulated containers into pipes driven into the ground. Liquid nitrogen is expensive, but the ground can be frozen much faster, which may justify the additional cost.

Table 3.1 Types of grout

Ground	Typical grouts used	Examples
Open gravels Gravels Coarse sands	Suspension	Cement suspensions. Cement clay, clay treated with reactants. Separated clay and reactants, montmorillonoid clays with sodium silicate and deflocculents (clay gels). Two-shot sodium silicate based systems for conferring strength. Bituminous emulsions with fillers and emulsion breaker.
Coarse sands Medium sands	Collodial solutions	Single-shot silicate based systems for strength. Single-shot lignin based grouts form moderate strength and impermeability. Silicate-metal salt single shot system, e.g. sodium silicate-sodium aluminate; sodium silicate-sodium bicarbonate. Water soluble precondensates. Oil based elastomers (high viscosity).
Fine sands silts	Solutions	Water soluble polysaccharides with metal salt to give insoluble precipitate. Water soluble acrylamide, water soluble phenoplasts.
Morainic	Any of the above	Any of the above. Choice of grout dependent on grain size and content of moraine.
Open jointed rocks; medium jointed rocks	Suspensions	Cement-sand, cement, cement clay.
Medium jointed rocks; fine jointed rocks	Solutions	Oil based elastomers, non-water soluble polyesters, epoxides and range of water soluble polymer systems given above. Hair cracks in concrete would be treated with a high strength low viscosity polyester or epoxy-resin.
Clays	Suspensions	Cement, cement clay.

Excavations

The methods of excavation and types of temporary or permanent support depend on a number of factors. These can include:

(i)   Type of sub-soil to be encountered;
(ii) The depth of excavation;
(iii) The existence of groundwater;
(iv) Type and extent of the excavation;
(v)   Whether the excavation is in close promixity to existing structures etc.

Excavations may be classified as: shallow, medium or deep excavations. The depth of each type is:

Shallow – Up to 1.5 m
Medium – Between 1.5 m and 6.0 m
Deep – Over 6.0 m

Most excavations for trenches and large pits is carried out using mechanical plant. The type of plant used will depend on the extent of the excavation and by the length of haul to the disposal point. Types and uses of excavating plant are discussed later in this chapter.

When conditions make it impractical to excavate mechanically, hand excavation becomes necessary. Such conditions include:

(a) Ground too steep for a machine or restricted working space;
(b) Very poor ground conditions;
(c) When the contract is too small for mechanical plant to be economical;
(d) Sites where underground services are known to exist.

In the case of small pits and shafts, methods of excavation are governed by the confined space and by obstructions caused by temporary supports. It may be necessary to use a combination of hand excavation and mechanical plant for removing the spoil from the area of the excavation.

Rock excavation
Excavation and breaking rock and other hard material will vary according to conditions, but could include:

(a) Use of pneumatic breakers;
(b) Use of hammers and wedge;
(c) Drilling with pneumatic machines and breaking by use of drawing plugs or freezing liquid;
(d) Drilling with pneumatic machines and breaking by blasting.

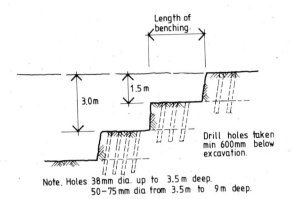

Note. Holes 38 mm dia. up to 3.5 m deep.
50–75 mm dia from 3.5 m to 9 m deep.

Figure 3.10   Rock excavation benching method

(a), (b) and (c) are used where any of the following conditions occur: could cause annoyance by blasting; adjacent buildings could be subject to damage; rock falls may result.

The two basic methods of drilling and blasting rock are:

> Benching
> Wellhole blasting

The benching method is best illustrated by referring to Figure 3.10.

Wellhole blasting involves drilling of holes 150 mm to 250 mm diameter at intervals, the spacing of which should be equivalent to the depth of face to be moved. The depth of boring should be at least 2 m and up to 24 m to be economical.

In both cases the holes are machine drilled to the required depth and loaded with a suitable explosive. This form produces good fragmentation and thereby reduces the amount of secondary blasting.

The cheapest and most popular form for blasting for more than a decade has been the use of ammonium nitrate. Its main disadvantage is its limited power particularly when hard rocks are involved.

A recent development in the explosives market has been the initiation of the explosives in the hole. Whereas a few years ago it was unthinkable to put a detonator down a deep hole, it has now become an accepted practice. The main advantage is that the explosives are being detonated at the bottom of the hole and the explosive wave is rising up the hole from the bottom to top. This gives an increase in fragmentation.

Methods of removing rock from excavations

The choice of method will depend on the position and size of the rock pieces, but they may be removed by the following means:

By hand loading on to flat bottomed rock skips;
By the use of mechanical excavators such as draglines, face shovels and loaders.

In both cases the material will be loaded into trucks, dumpers or railway wagons.

Support of excavations

Timber, steel sections, or precast concrete can be used as temporary supports for excavations.

The selection of the method of supports is largely influenced by the type of ground encountered, but the plant used for excavation and the amount of re-use possible of the temporary supports are also factors to be considered.

To maintain safety standards when working in excavations, Part IV of the Construction (General Provisions) Regulations requires that an adequate supply of timber or other support material be provided and used in excavations of more than 1.25 m in depth. This requirement is to prevent danger to any person working within an excavation from a fall or dislodgement of earth, rock or any other material forming the sides of the excavation. Further requirements of this Regulation are that:

(a) A competent person should inspect the excavation at the start of each day's work or at the beginning of each shift.

(b) A thorough examination be carried out after the use of explosive charges, damage caused to, or collapse of, any part of the excavation and support work and, in any case, every seven days.

A written report giving full details of the examination including the name of the person who carried it out and the date must be made.

(c) All materials to be used must be inspected before their use.

(d) The support work must be properly constructed and struts and braces securely fixed against accidental dislodgement.

(e) If there is a risk of the excavation being flooded, ladders or other means of escape must be provided.

(f) If an existing structure is likely to be affected by the close proximity of an excavation, that structure must be shored or supported against collapse.

(g) Excavations of more than 2 m in depth which are close to persons working at ground level or passing by, must be fenced off with guardrails.

When access to the excavation is required by plant or the movement of materials the guardrails may be temporarily removed but must be replaced as quickly as possible.

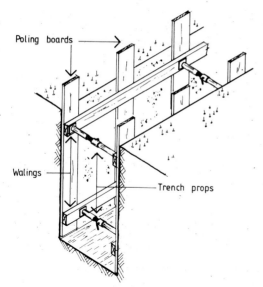

Figure 3.11 Support for shallow trench in firm ground

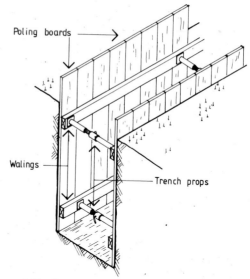

Figure 3.12 Support for shallow trench in loose ground

(h) To avoid the collapse of the side of an excavation or materials falling on to persons working within the excavation, plant materials and spoil must be kept clear of the edges.

The fundamental requirements of support work may be summarised as follows:

(a) To provide safe working conditions;
(b) To allow the efficient execution of both the excavation and permanent construction work;
(c) Be capable of being easily and safely removed after completion of the permanent work.

Temporary support for trench excavations

For shallow trenches in firm ground it is usually only necessary to provide simple strutting to the sides of the excavation. This method is illustrated in Figure 3.11 which shows the position of the poling boards and waling which are kept in position by adjustable steel trench props at 600 mm centres approximately. When excavating in loose ground it is necessary to provide poling boards close together to prevent the soil from collapsing into the trench, thus endangering operatives who may be working in the area, see Figure 3.12.

When medium trenches are required, it is advisable to use close poling with additional walings and trench props, but in the case of very loose or waterlogged ground, support must be provided before excavation work commences. This can be achieved by driving timber runners or steel sheet piling to a depth below the formation level, or by a 'dig and drive' operation, in

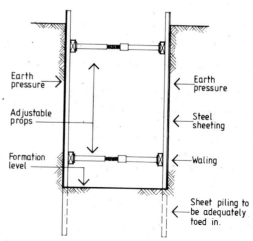

Figure 3.13 Support for medium depth trench using sheet piling

which the runners are driven to a depth of 1.5 m approximately, followed by the excavation of the trench, and the operation is then repeated until the required depth has been reached, see Figure 3.13.

For deep trenches, a system of steel sheet piling is suggested, supported by puncheons, walings and trench props, for details see Figure 3.14.

When ground conditions permit, it may be advantageous to use a moving proprietary shoring system. This method consists of sheeting panels which can be lowered into a trench and jacked against the sides. The whole assembly of sheeting and supports can be moved along the trench as work proceeds.

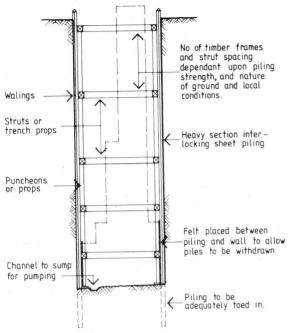

Walings

Struts or
trench props

Puncheons
or props

Channel to sump
for pumping

No of timber frames
and strut spacing
dependant upon piling
strength, and nature
of ground and local
conditions.

Heavy section inter-
locking sheet piling

Felt placed between
piling and wall to allow
piles to be withdrawn

Piling to be
adequately toed in.

Figure 3.14   Support for deep trench using sheet piling

Figure 3.15   Supports for deep pit excavations

Temporary support for pits and shafts

Sides of pits and shafts are supported by poling boards, horizontal sheeting or steel sheet piling. These supports can be held in position by horizontal struts, ground anchors or by raking shores.

For large deep pits, driven interlocking steel sheet piling is normally used supported by walings, which can be either timber or steel channel sections, see Figure 3.15.

When shallow pits are to be excavated, sheet trench sheeting or closely spaced timber poling boards are used supported by walings.

Support for the vertical members can be achieved by the use of horizontal strutting. The spacing and size of the struts will depend on the force to be resisted, see Figure 3.16. Structurally this method is very efficient, but it does have the disadvantage of struts obstructing the construction work and also limits the choice of excavating plant.

On very wide excavations, where room permits, the sides may be supported by raking shores, see Figure 3.17, but this method has, to some extent, been replaced by anchored systems of support. This method has the advantage of providing a working area free from obstructions such as rakers and horizontal struts.

Steel sheet piling is driven to form a curtain around the area to be excavated. When the excavation work starts ground anchors are used to tie back the sheet piling.

Ground anchors consist of bars or strands and are stressed against an anchorage which holds the tie in the ground. A tensioning anchor is provided at the other end. The ground anchor is positioned by boring a hole using a rotary percussion drill with casing in sand or gravel. An auger is used in clay soils. The bar or strands are then inserted and a predetermined length grouted up with neat cement, by pressure. When the anchorage has gained its strength the bar or strands are then tensioned to the required force. Figure 3.18 illustrates this method.

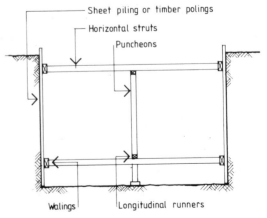

Sheet piling or timber polings

Horizontal struts

Puncheons

Walings

Longitudinal runners

Figure 3.16   Support for shallow pits using sheet piling and timber polings

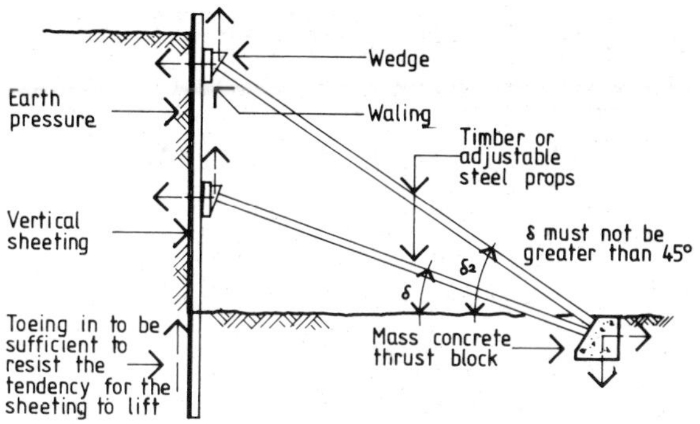

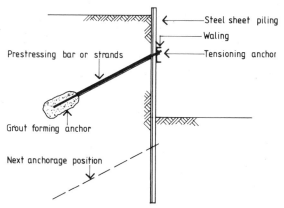

Figure 3.17 Shoring to wide excavations

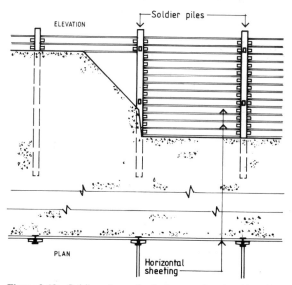

Figure 3.18 Anchored sheet piling to excavations

Another method of support which differs from previous types is that the wall of the excavation is supported by sheeting that spans horizontally between soldier piles, see Figure 3.19.

Soldier piles are usually H section steel piles driven at predetermined centres before excavation work is commenced. As the ground is excavated the horizontal sheeting (it may be timber, trench sheeting or pre-cast planks) is lowered down and wedged tight between the soldiers.

A new method based on this system uses steel panels between the soldiers that lower themselves due to their self weight. The soldiers are normally strutted in trench excavations or anchored back in open excavations.

Permanent supports for excavations

When it is necessary to construct an underground structure (such as a basement) on a very confined site, where insufficient space is available for conventional temporary supports, then a method of permanent support should be considered.

This permanent support may take the form of diaphragm walls, contiguous bored piles or interlocking bored and cast-in-place concrete piles. When ground movement has to be reduced to a minimum because of the close proximity of an existing structure, this type of support should be considered.

There are several construction methods available. When diaphragm walls are used, they should be constructed as suggested in the earlier part of this chapter. Rigid strutting can be provided to the wall by the

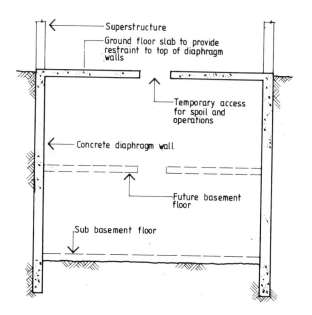

Figure 3.20 Permanent support for excavations (1)

Figure 3.19 Soldier pile method of supporting the sides of excavations

ground floor and any lower suspended floor. Excavation can then take place, the spoil being removed through openings left in the floor(s) for this purpose, see Figure 3.20.

If the permanent support is to be provided by piling, then it may be necessary to use steel lattice beams at the top of the piles to give horizontal temporary support in one or both directions.

Excavation, and then subsequent construction of the basement can take place, see Figure 3.21.

Open excavations

Providing ground conditions are suitable and there are no structures in the close proximity, it may be possible to cut back the perimeter of the excavation to the natural angle of repose, thus avoiding the necessity of supports.

With this type of excavation, it should be inspected regularly for local earth slips resulting from marginally unstable conditions. If such local slips do occur there may be concern for the safety of operatives working in the excavation, or possible damage to the partly built works.

If an open excavation is to be of a permanent nature, as in the case of a motorway cutting, the angle of the slopes should be considered in relation to the appearance of the cutting in its surroundings. It may, of course, be necessary to vary the slope angles of the cutting.

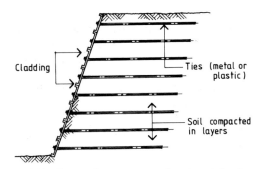

Figure 3.22   Support for open excavations by reinforcing retained earth

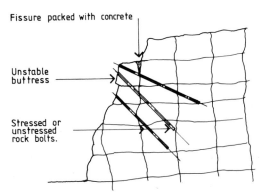

Figure 3.23   Unstable rock slopes stabilised by bolts

When a steep slope is required, it may be possible to reinforce the retained soil by using horizontal metal or plastic ties. The face of the excavation is protected by steel or reinforced concrete cladding elements, see Figure 3.22.

To prevent slipping of some rock faces towards an excavation or cutting, prestressed concrete anchor bolts can be grouted into holes drilled deep into sound rock. Rock bolts can be of considerable help in preventing falls from a steeply cut rock face. When a good appearance is not required, the slope can be covered with wire mesh pinned to the rock face by short bolts, see Figure 3.23.

Further information on excavations is given in BS 6031 *Earthworks*.

Cofferdams

A cofferdam is a temporary structure built for the purpose of excluding water or soil from an excavation and permitting construction to progress without excessive pumping and, in the case of land cofferdams, to support the surrounding ground.

In selecting a cofferdam type for a particular contract, the following conditions should be taken into account:

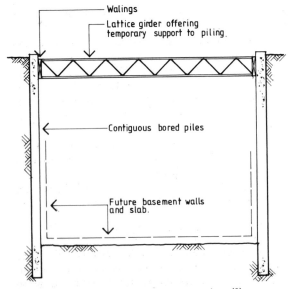

Figure 3.21   Permanent support for excavations (2)

(a) Site investigation report;
(b) Whether a land or water cofferdam is required;
(c) The nature of the permanent works to be built;
(d) The dimensions of the working area required inside the cofferdam;
(e) The depth of water or earth to be retained;
(f) Soil conditions;
(g) Construction time;
(h) Methods of constructing and dismantling cofferdam;
(i) Proximity to existing structures.

Cofferdams can be divided into the following types:

(i) Earth and rock fill cofferdams;
(ii) Crib cofferdams;
(iii) Double skin cofferdams;
(iv) Diaphragm wall cofferdams;
(v) Single-skin sheet pile cofferdams.

(i) Earth and rock-fill cofferdams

This type consists of a dam of earth or rock constructed usually during a low water period. Because of their temporary nature, precautions against leakage are less rigorously enforced, although construction should be in accordance with the principles used for permanent works. The upstream face of the dam should be protected by rubble or concrete although other materials can be used.

(ii) Crib cofferdams

Construction of this type of cofferdam can be in either timber or pre-cast concrete units. The units are framed up in a crisscross fashion and are connected together by bolts and timber connections to form pockets 3.5 m square. When located in position the cofferdam is filled with rocks and finer material to minimise infiltration. The water face may be lined with sheet piling to make the dam more watertight, see Figure 3.24.

(iii) Double skin cofferdams

Double skin cofferdams are self supporting gravity structures either of the parallel sides double wall or cellular type.

Double wall cofferdams usually consist of two rows of sheet piling driven into the ground and tied together at the top by walings and tie bolts, see Figure 3.25. The space between the rows of piles is filled with either sand, gravel, crushed rock or broken brick. Weepholes have to be provided near the bottom of the piles on the inner side so as to reduce the pressure on the inner line of piling. Cofferdams of this type are not recommended for land excavations, since it is more economic to use a

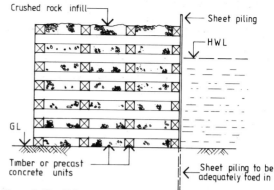

Figure 3.24 Crib cofferdam

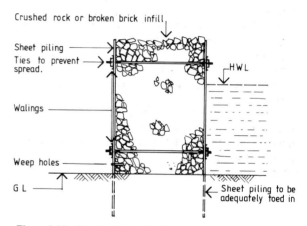

Figure 3.25 Double skin cofferdam

single row of piling supported at the top by rakers or other forms of bracing.

Cellular cofferdams are used for work in rivers and the sea. They consist of complete circles of interlocking piles. Trough shaped sections are not suitable for this type of work, because the interlocks will open out. Care should be taken with the pile driving, as a major failure at an interlock may result in the failure of the dam.

Cell filling can consist of sand, gravel, crushed stone or broken brick and drainage similar to that used for the double wall type will be required.

An adequate factor of safety must be allowed for against tilting and sliding during the design stage.

(iv) Diaphragm wall cofferdams

This form of cofferdam may be constructed by surrounding the perimeter of the excavation with a continuous cast-insitu diaphragm wall as described in the earlier part of this chapter.

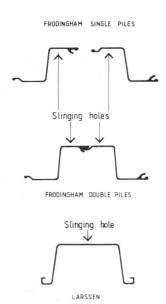

Figure 3.26   Steel sheet piling

(v) Single-skin sheet pile cofferdams

Sheet piling is widely used for cofferdams. Patent steel sheet piling such as Frodingham and Larssen is the most common form because of its structural strength, the watertightness provided by interlocking sections,

and its ability to be driven into most types of ground.

Sheeting without bracing should be designed as cantilevers fixed in the ground, but for deep excavations it may be necessary to restrain the top of the sheeting as described in the earlier part of this chapter.

Timber sheet piling has been used for this form of works and is still used in countries where timber is plentiful but, due to the high cost, is now rarely used in the United Kingdom.

Sheet piling

Sheet piling may be generally defined as closely set piles driven into the ground to keep earth or water out of an excavation.

There are three basic types of sheet piling:

1   Steel sheet piles;
2   Reinforced or pre-stressed concrete sheet piles;
3   Timber sheet piles.

Steel sheet piles

This type of piling is by far the commonest form of sheet piling used in civil engineering works today. It is used for the construction of permanent retaining walls in docks, harbour and embankments, river bank and sea defence works. Steel sheet piling is also used to support the sides of trenches and open excavations and forming cofferdams.

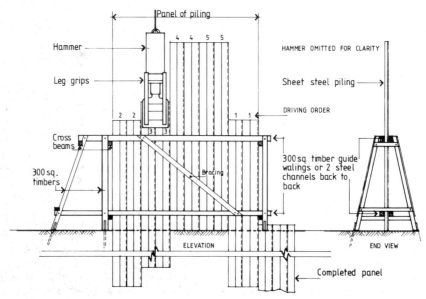

Figure 3.27   Typical piling trestle

Reinforced concrete and prestressed concrete sheet piles

Generally used for permanent work that will be incorporated into the completed structure.

Timber sheet piles

Due to the cost, timber sheet piling is rarely used in the United Kingdom, but in countries where timber is plentiful and steel or precast concrete sheet piling is not readily available, it is used quite extensively. In such areas, timber sheet piling is mainly used for temporary structures in a similar fashion to steel sheet piling.

Construction methods

Construction methods of driving sheet piling are described below:

Steel sheet piles

The two main types of steel sheet piles used in the United Kingdom are 'Frodingham' and 'Larssen'. Although similar in application of use their difference is the position of interlock. When in position, both systems produce a wall which is trough-shaped in plan, see Figure 3.26. Frodingham piles are available in lengths of 9.0 m to 24 m and Larssen in lengths of 9 m to 26 m. In special cases the piles can be supplied in lengths of up to 30 m though handling such lengths of pile may prove difficult. As an alternative to using such long lengths or in cases where the length of pile required has been underestimated, the pile may be increased in length by splicing on an extra length of pile with fishplates or by the use of site welding.

To keep the piles in position during driving, it is essential to erect temporary guide walings. These can either be of trestle construction, or formed by the use of timber or steel piles, see Figure 3.27. In general, the choice of timber or steel to form these temporary works will depend on relative cost and availability.

The piles should be supported or guided at two levels during driving by the walings, one pair being set as near to ground level as possible or in the case of marine works, just above the low water mark. To assist stability of the driving frame the distance apart of the two sets of walings should be as large as is practicable. It is of the utmost importance that the walings be held rigidly in position. The vertical supports to the walings should be set at about 2.5 m or 4 m centres.

Sheet piling may be driven by drop, steam, air or diesel hammer. The hammer is normally suspended from a derrick or crane and fitted to the piles by means of leg-grips or leg-guides to obtain correct positioning of the hammer on to the piles.

In certain circumstances, such as in urban areas, or where the work is close to an existing structure, the noise and vibration caused by the action of the hammer may be unacceptable. In such cases an alternative method of driving is used.

The method most commonly adopted is to use hydraulic rams to push the piles into position. The Taywood Pilemaster, which is used throughout the United Kingdom and abroad, is a machine that works on this principle. Fitted with eight hydraulic rams, the Pilemaster is capable of driving and extracting sheet piles with little or no measurable vibration and very low noise (69dBA at a distance of 1.5 m from the machine) – see Figure 3.28. The most commonly used method of driving steel sheet piles is known as 'driving in panels'. This method is carried out in the following stages:

Figure 3.28 Taywood Pilemaster

(a) A pair of interlocked piles are pitched and part driven to a depth level with the top waling and carefully checked for verticality.

(b) About six to twelve pairs of piles are then pitched and interlocked in position and held securely in the guide walings. Hence a panel of piles is formed, the first pair being partly driven.

(c) The hammer is now transferred to the last pair of undriven piles in the panel which are then part driven to about the same depth of the first pair. Again these must be checked for verticality.

(d) The remaining pairs of piles are then driven to their final level, starting with the pair adjacent to the last and working across to the pair adjacent to the first.

(e) The first pair is not fully driven as they will form the first guide pair of the next panel. The procedure is then repeated.

Reinforced and prestressed concrete sheet piles

Generally used for permanent work that will be incorporated into the completed structure. The cross section of this type of pile is generally rectangular with either tongue and grooved or 'birdsmouth' shaped sides to form the interlock. The foot of the pile, as shown in Figure 3.29 is bevelled and feather edged to assist in the

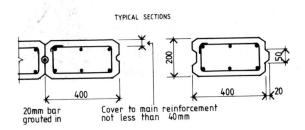

TYPICAL SECTIONS

20mm bar grouted in

Cover to main reinforcement not less than 40mm

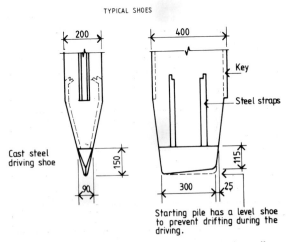

TYPICAL SHOES

Cast steel driving shoe

Key

Steel straps

Starting pile has a level shoe to prevent drifting during the driving.

Figure 3.29 Typical precast reinforced concrete sheet pile details

Mild steel hoops are placed around the top of the pile to prevent spread during piling; in soft ground conditions pile toe is bevelled and pointed, and tied with m.s. hoops as for the toe; other ground conditions require a cast steel point attached to the toe.

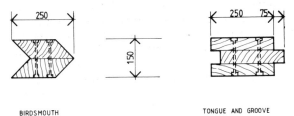

BIRDSMOUTH   TONGUE AND GROOVE

Figure 3.30 Typical timber sheet piles

driving and also to help correct a tendency for this type of pile to lean (or drift) as it is driven.

When the pile is to be driven in hard strata, a metal shoe is used on the foot to protect it during driving. As it has been noted, there is a tendency for this type of pile to drift as it is driven: to counteract this problem the piles are driven one at a time working from left to right. The first pile must be plumb, of course, and careful control is required during the driving of it, the following piles will, as they are being driven, tend to drift towards the first pile hence forming a close fit. If the completed work is to be watertight the joints are usually grouted up.

Timber sheet piles

The simplest form of timber sheet piling is rectangular planks mainly used as close boarding in trench excavations. Other shapes include birdsmouth (formed by bolting together double bevelled planks) and tongue and grooved, (formed by bolting together three rectangular planks).

Timber sheet piles are driven in a similar manner to RC and pre-stressed concrete sheet piles, see Figure 3.30.

Sheet piling

Sheet piling is extracted by the use of a double-acting hammer in which the motion is reversed, or they can be jacked out. The Taywood Pilemaster extracts piles by reversing the hydraulic rams.

Earthmoving plant

Most civil engineering contracts by the very nature of the work contain some element of earthmoving. It may form a fairly minor part such as stripping a site of top soil and excavating for pad foundations or be a major element such as in the construction of an earthfill dam. Whatever the size of the earthmoving content of the contract, one fact common to all is that the contractor will employ mechanical aids to carry out the work.

It is important from the outset when considering the selection of an earthmoving system that the basic roles and limitations of the various types of excavating, loading and hauling plant available are clearly understood. This is necessary, as the first stage in the selection of an earthmoving system is to make a preliminary selection of practical alternatives based on the type of excavations to be carried out, the conditions of the site, the contract period and the availability of any one type of machine.

Only after a comparison of the alternatives, based on calculated production rates and cost performance of each system plus practical experience, can the final selection be made.

Types of earthmoving plant

Scrapers

The principle of the scraper is that an open bowl, which has at its base a blade that can be raised or lowered, is drawn along and by lowering the blade, which 'cuts' into the ground as the scraper moves, the material over which it passes is scraped up into the bowl. When the bowl is full the blade is raised and the scraper can move

Figure 3.31 Elevating scraper

Table 3.2 Selection of scrapers

Scraper Type	Material operating range						
	Clay	Silt	Sand	Gravel	Well broken rock	Ripped rock	Blasted rock
Conventional Elevating						··········	
Tandem powered (a) Conventional (b) Elevating				···········	···········		
Dual 'push-pull'	···				····		

Applicable _____
Marginal ....

Figure 3.32   Bulldozer acting as a pusher on back of scraper

Figure 3.33   Tandem powered scraper

to the dump area where, by lowering the blade sufficiently, the earth held in the bowl is pushed out as the scraper moves. It can be appreciated that the scraper is a very important machine as it can excavate, haul and dump, and is the only earthmoving machine that is specifically designed to do all three.

There are, of course, operating limitations to the scraper, the two main considerations being the length of haul, that is the distance from the cut to the dump area, and the type of material being excavated.

As a general guide, if the haul distance is less than 100 m or greater than 2500 m then an alternative system such as dozers, loaders and trucks may well prove more economic. Rock either blasted or ripped is not in most cases a suitable material for scraping. Table 3.2 can be used as a guide to the selection of scraper type according to material. Another factor to be considered is that even if the scraper can handle well broken rock, from the point of excavation, excessive wear on the tyres may preclude the economic use of a scraper.

Types of scrapers

Elevating scrapers (Figure 3.31)

Elevating scrapers have in the bowl a mechanism known as the elevator that assists the excavated material into the bowl. These machines are self loading and do not require the assistance of a pusher for loading. The advantage of the elevating scraper is that (a) the loading time is shorter, and (b) additional plant, i.e. a crawler dozer is not required. However, the elevator adds to the operating weight of the machine which in turn over long haul distances is a disadvantage and the overall running and operating costs tend to be higher than that of the conventional scrapers.

Conventional wheel tractor scraper (Figure 3.32)

This type of scraper is what one might call the basic model. It has the lowest owner and operating costs of the various forms of scraper, but requires the assistance of a crawler dozer when loading. Powered by a single engine wheel mounted tractor unit, it has the widest operating range of materials and is the normal choice where long hauls on well maintained and fairly level haul roads are in use.

Tandem powered units (Figure 3.33)

Both conventional and elevating scrapers are available with twin power units, one at the front and one at the rear. Because of the front wheel drive and, of course, the increased power, these units can operate in severe conditions both in the cut and on haul. Where gradients are particularly steep, the tandem powered scraper may be the only practical choice for efficient operating.

'Dual load' or 'push pull'

A technique, rather than the type of scraper, known as dual load or push pull is a method often used to great advantage. The principle is that two tandem powered scrapers are coupled together in the cut area combining their power and tractive effort and hence dispensing with the need for pusher assistance from a crawler dozer. When fully loaded the machines uncouple and haul their loads in the normal manner.

Tractor towed scraper

Designed for use on moderate to low volume earthmoving projects this form of scraper is as the name suggests a scraper bowl towed by either a wheel or track mounted power unit. Typical applications are site preparation such as top soil stripping on commercial sites and minor road works such as improvements etc. When towed by a wheel mounted unit it has the advantage of high speeds over haul distances of 50 to 200 m but its tractive effort will be low. The scraper bowl unit can be either conventional in design and hence require dozer assistance when loading, or self loading with an elevator mechanism in the bowl.

Crawler dozer (Figure 3.34)

Crawler dozers are powerful track mounted units which are used for a variety of work on earthmoving projects. Their high tractive ability in soft conditions enable them to work on ground that would defeat wheel mounted vehicles, and they can also operate as an economic alternative to scrapers on short hauls of up to 90 to 100 m. The selection of the correct dozing tool or blade is important to achieve maximum productivity. The main types of blade in use are:

Figure 3.34 Crawler dozer

'U' blade

So called because it has large 'wings' at the edges of the blade to retain the load, the full universal blade is designed for moving loose materials over large areas. It must not, however, be used for push loading scrapers as the 'wings' can damage the rear tyres of the scraper.

Semi 'U' and 'S' blades

Semi universal and straight blades are used for most production dozing, i.e. land clearing, excavations, grading and stripping. The actual choice of either series 'U' or 'S' blades tends to be one of availability and personal preference. The straight blade is to be preferred if the dozer is required to push load scrapers. Both blades should, however, be fitted with a push plate to avoid damage when used for this work.

'C' blade

The cushion blade is specifically designed for push loading scrapers on-the-go. Rubber cushions allow the dozer to absorb the impact of contacting with the scraper push block. When not pushing scrapers the 'C' blade equipped dozer can carry out general cut maintenance.

'A' blade

The angle blade is a straight blade that can be turned to either side through about 25° and is mainly used for backfilling of trenches, side casting and contouring work.

Rear accessories (Figure 3.35)

The main rear accessories used on crawler dozers is the ripper. This is used for ripping up material that is too hard for normal excavating methods, prior to excavations. There are three basic types of ripper:

(i)   Radial
(ii) Parallelogram
(iii) Variable pitch

  (i) The radial ripper has a beam with the shank fixed at one end and pivoted to the rear of the dozer. The beam can move through a vertical arc of about 30° and generally has provisions for up to four shanks. The angle of the shank to the beam can be adjusted by the insertion of pins, but cannot be adjusted during ripping which means that the angle of the ripper constantly varies with depth. One advantage of this system over the parallelogram system is that when the ripper is lifted out it moves back along its arc reducing the problem of trapping the shank through rock catching between the shank and the dozer.

  (ii) With the parallelogram ripper the linkage carrying the beam and shank maintains the tip at a constant angle irrespective of depth.

  (iii) The variable pitch ripper combines the benefits of the radial and the parallelogram rippers. It has the added advantage of being able to vary the tip angle during ripping to obtain optimum effect.

The number of shanks used will depend on the ripability of the material. The harder the ripping the greater the power required and hence in very tough material, only one shank is used as all the power is applied to the one tooth.

A further rear attachment is the scarifier which is similar to a ripper but is lighter in construction and used for breaking up black-top surfaces etc., prior to dozing.

Loaders

Wheel loaders (Figure 3.36)

The wheel loader is designed for top loading of haul units. It can handle most materials and, if necessary, haul up to about 100 m economically, eliminating the need for haul units. Most wheel loaders are steered by an articulated system. This type of steering permits the loader to maintain a long wheel base for stability while keeping a small turning radius. The machine literally pivots in the middle. This system of steering allows it to work with great speed in fairly confined areas. The disadvantage of the loader is that it cannot work with material that is below its standing level.

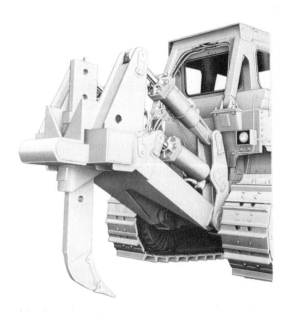

Figure 3.35   Ripper attachment

Figure 3.36 Articulated wheel excavator/loader loading an off-highway dump truck

Crawler loaders

This type of loader only differs from the wheel loader in that it is mounted on tracks instead of wheels and is therefore not articulated. It has, in recent years, declined in popularity as it is apparently more difficult to operate than the equivalent articulated wheel loader. However, in difficult ground conditions where high tractive effort is required, the crawler version will operate more efficiently than the wheeled mounted version.

Bucket selection

Bucket capacities range from 0.7 m³ up to 10 m³. The selection is based on the density of the material being loaded (it is important not to specify a bucket that if fully loaded would upset the stability of the loader) and the physical state of the material.

Bucket types

General purposes

This bucket, which has a straight lip with or without teeth, is designed for loading material which if the bucket is not equipped with teeth can be broken loose with relative ease, or with teeth requires a certain degree of penetration and breakout force. Typical materials handled would be sand, gravel, soil and clay.

Multi purpose bucket

A very versatile bucket that in soils can be used as an ordinary bucket, a grab bucket, a dozer blade or grader. It can also be used for picking up pipes and irregular shaped objects.

Vee lip bucket

Known also as a rock bucket, the cutting edge is 'V' shaped and the sides low cut to enable large pieces of material to be handled with ease. Used for loading blasted or ripped rock.

Side dump bucket

Used as the general purposes bucket, the difference being that it tips sideways instead of forwards. This can reduce loading time and in confined working spaces aid close quarter loading.

Figure 3.37   Track mounted hydraulic hoe excavator

Hydraulic excavators (Figure 3.37)

For the excavation of trenches and shafts the hydraulic excavator which on smaller machines can be wheel mounted but in general is track mounted, has little real competition for efficiency of operation.

The basic principle of the excavator is a hydraulically controlled one or two part boom with a dipper stick attachment on to which the bucket or clamshell is pivoted. The boom and operator's cabin are set on a rotating base enabling the machine to excavate and load, whether wheel or track mounted, without moving its position. It is capable of excavating in most materials, except continuous rock, at depths of up to 10 to 12 m for the larger machines. In general, two types of bucket attachment are available:

(i)   Backhoe or excavating bucket
(ii) Clamshell bucket

(i) Backhoe (Figure 3.38)

The backhoe bucket is for trenching and similar work. Because of the hydraulic power system used to operate the bucket, the backhoe has a very high penetration and breakout force with a maximum depth of cut of 10 m on the largest machines. When selecting a bucket the width and bucket tip radius are important factors. In general, in easily excavated materials the widest bucket, with respect to the trench width, will achieve the best production, but, in difficult material, a narrow bucket with a short tip radius will prove most effective.

Figure 3.38   Backhoe bucket

(ii) Clamshell bucket (Figure 3.39)

The clamshell is an important attachment as it gives the excavator the ability to work on vertical shafts and in confined areas where existing services pipes would impede the backhoe. With full dipper extensions, depths of up to 12 m are within the capability of the larger machines.

Hydraulic face shovel (Figure 3.40)

Similar in principle to the hydraulic excavator but, as will be noted in Figure 3.40, the digging action is forward and therefore is suited to excavating not only below its standing level, but above. This ability makes it suitable for large basement and similar excavations.

Figure 3.39 Clamshell bucket

Again it is a very powerful excavator and is often used in quarry work. Two types of bucket can be used which are similar to the general purpose and multipurpose buckets described in the loader section.

Haul units

Where plant other than scrapers are used for excavating haul, units must be available to transfer the material to the dump area if it is not close to the excavation. The majority of haul units used on large contracts tend to be 'off-highway', that is because of their size and axial loads when fully loaded, they are not allowed to travel on public highways. Further 'on highway' trucks do not usually have the robustness required in their suspension and general construction to be used over long periods on earthmoving contracts, and should only be used if the dump area can only be approached over public roads. 'Off-highway' haul units

Figure 3.40 Hydraulic face shovel

Figure 3.41   Rear dump

can be generally classified by their method of dumping, the two types being:

(i)   Rear dump
(ii) Bottom dump

(i) Rear dump

Rear dump trucks are by far the most popular form of haul units used on earthmoving projects. They are extremely tough vehicles with capacities of up to 317,500 kg and can haul any material. The two types most commonly seen are the traditional rigid form (Figure 3.40) and, to a lesser extent at present, the articulated form (Figure 3.41). The articulated rear dump truck is a relatively new addition to the haul unit range, but because of such features as excellent manoeuvrability, bogie action rear wheels on some models which allow them to haul over rough unprepared haul routes, and high tractive power, they are sure to gain in use.

(ii) Bottom dump

Bottom dump haulers are used for hauling free flowing materials over long distances where the dumped material is to be spread. Both rear and bottom dump haulers can be towed by wheel or track mounted tractor units. The advantage of this type of haul unit is that the tractor can be detached from the unit and used to tow a scraper bowl, when it is not required to tow the haul unit.

Dragline (Figure 3.42)

In certain conditions the dragline excavator is really the only practical choice of machine available, and should be considered as a specialist excavator.

The machine is normally track mounted with the bucket suspended from a crane jib. The machine excavates by casting out the bucket and then dragging it back. Although the very large draglines used in quarry work can deal with blasted rock in the construction world, it is used for working on very soft ground as in rivers, the machine operating from the bank or off a pontoon. The bucket capacity ranges from 0.5–4.5 cubic metres. The main disadvantage of the dragline is that a considerable amount of skill is required on the part of the operator, both in excavating and loading, to ensure good productivity. However, as noted above, its specialised applications make it a most useful excavator in the right conditions.

Motor grader

A wheel mounted articulated machine with the blade mounted midway between the front wheels and the rear drive wheels. It is used for spreading, levelling and grading. The blade is angled to allow the excess material to be pushed to one side as the machine

Figure 3.42   Crawler dragline

moves. The front axle and wheels are usually capable of turning through 50°, oscillating vertically and leaning (Figure 3.43) allowing the machine to remain stable on uneven and inclined surfaces. Another feature of the motor grader is the use of a rear mounted set of scarifiers for breaking up tarmacadam road surfaces and light concrete slabs etc., prior to removal.

Haul roads

It will be clear that the haul route and its surface condition can have a considerable effect on the total cycle time of an earthmoving operation and, in many cases, may be the major part of the cycle. The route the haul road follows, its surface and maintenance programme, are important factors to consider when planning an earthmoving project. As with so many elements of earthmoving work, the decisions finally taken must be based on a balance between that which is desirable in theory and that which is obtainable in terms of the site conditions and, of course, the economic restraints of the contract.

The route – factors that reduce the operating speed of the plant using the road such as steep inclines, sharp bends, restricted crossing points etc., should where possible be avoided. The shortest route from the place of excavation, the cut, to the dump area may not always be the most efficient in terms of travel time for heavy earthmoving plant.

The road – on large earthmoving projects it is often necessary to construct the haul roads, this may take the form of anything from a graded earth road to a full metalled surface. Whatever the construction, it should be of a form that will allow all weather use, and provide a surface suitable for the plant using it without the need of a maintenance programme whose cost may outweigh any savings made in the choice of construction for the road.

Figure 3.43   Motor grader

Maintenance – although earthmoving plant is in general robust in design, a poorly maintained haul road will soon begin to take its toll in damaged axles and tyres and, of course, could prove hazardous. A planned maintenance routine should be established to maintain the haul road surface, and this is usually carried out by graders working stretches of the road of about a kilometre in length. Care must be taken though not to create a hazard by obstructing the passage of the haul vehicles with maintenance plant. Dust may also be a problem and as suggested in the section on safety, damping down may be necessary. Damping down can be achieved by spraying with water or by applying a coat of bitumen spray.

Safety and the operating of earthmoving plant

Earthmoving machines and the locations in which they operate must at all times be considered as potentially hazardous. The physical shape and size of the machines inevitably means that the driver's side and rear vision will be limited, also the noise of the engines would, in most cases, render shouted warnings inaudible to the driver. The machines are, of course, very powerful, a carelessly swung bucket or manoeuvre is only too often the cause of serious accidents.

Although the existing safety regulations make few specific requirements as to the safe operation of earthmoving plant, their general requirements for the provision of safe working places and procedures are, of course, applicable to this area of civil engineering work. The main legislation and codes of practice are listed below and should be referred to and studied.

Health and Safety at Work etc., Act 1974
Construction (General Provisions) Regulations 1966
Construction (Lifting Operations) Regulations 1961
BS6031: 1981. *Code of Practice for Earthworks*
Highways Act: 1959

The following list of safety precautions including legally enforceable requirements and procedures that although not spelt out in the relevant regulations are considered to be within their meaning and spirit, and therefore essential to the interests of safe working procedures with earthmoving plant.

Only trained personnel over the age of 18 years may operate earthmoving plant.
Unless seating is provided for the purpose, site personnel must not under any circumstances 'hitch' rides on the machines.
The operator must never leave the machine with the engines running, or while it is moving.
All required guards and shields must be in place when the machine is in operation.

A roll over protection structure (ROPS canopy) must be fitted to crawler and wheel mounted loaders and tractors, scrapers and graders, also when necessary a falling objects protection structure (FOPS canopy). The machine must be regularly maintained by trained mechanics, and inspected at least every seven days while operating.

When parked the machines must be left in a safe position, that is for example with loaders the bucket must be resting on the ground etc.

Loading buckets and haul units must not be overloaded. Haul units should be fitted with reversing horns as a warning to personnel.

Haul routes should be well maintained and, where necessary, warnings posted to warn of heavy earth-moving plant.

Haul units must not be driven at speeds above that laid down. If dust is created by traffic along a haul route that could cause a hazard through reduced visibility water bowsers should be employed to damp down the surface.

Where dumpers or haul units are tipping over the edge of a tip or excavation, stop logs must be provided to prevent the machine from moving too close to that edge.

When working on slopes crawlers and scrapers should work with the gradient and not across.

Noise

The noise produced by plant involved in earthmoving work is, more often than not, very loud. It can, if precautions are not taken, damage the hearing of the operators and those working in close proximity. In populated areas it can be a nuisance to those who live or work near the site or haul roads. It is important to give due consideration to this problem at the planning stage of the project to avoid injury (impaired hearing) to site operatives and to avoid any unplanned delays in the contract if, for example, a local authority enforces restrictions on the hours of work or type of plant to be used.

There is a considerable body of legislation controlling noise, the main Acts being:

The Control of Pollution Act 1974
The Public Health Act 1961
The Health and Safety at Work etc. Act 1974

The Control of Pollution Act 1974. Section 60 makes provision for local authorities to control the noise levels emanating from a construction site in order to protect people in the locality of the site from excess noise levels. As a precaution the contractor, before commencing work, should, if considered necessary, make an application to the local authority to carry out the work and include in the application a description of the pro-

Table 3.3 Maximum exposure to noise levels
(Health & Safety Executive)

Exposure duration hours per day	Maximum sound level level in dbA
8	90
4	93
2	96
1	99
½	102
¼	105

posed works and any precautions to be taken to minimise the noise.

For further guidance on this subject BS 5228: 1975 *Code of Practice for Noise Control on Construction and Demolition Sites* should be consulted.

Noise control

The Health and Safety Executive have issued as a guide a table of continuous and intermittent noise levels, which unless hearing protection is worn, would constitute a serious hazard to hearing. Refer to Table 3.3.

It is estimated that a 55 h.p. bulldozer will produce, when operating, a sound level somewhere between 101 to 108 dBA or a scraper up to 128 dBA. It is necessary then that the operator of such earthmoving plant and any persons exposed to its noise and working in close proximity, should wear suitable ear protectors. Also, to reduce the noise level at source, the silencers on the exhaust system should be checked for maximum efficiency and any enclosure panels over the engines kept closed.

Noise is an unavoidable consequence of excavation work but steps can be taken at all stages to reduce the level of noise spreading from the site. Screening of the work can make a very useful contribution to this often difficult problem. The screen, which can either be formed out of continuous timber boarding or earth mounds for example, should for maximum effect, be positioned as close as possible to the area from which

POORLY GRADED FILL   WELL GRADED FILL

Figure 3.44 Fills

the noise is emanating. Static plant, such as compressors and pumps can, if warranted, be enclosed in an acoustic shed but, as carbon monoxide fumes are lethal, provision must be made for good ventilation. If the noise is of a fluctuating nature, for example the detonation of explosives, the contractor should, where possible, inform the public of the times of the explosions. This can make a significant contribution to the acceptance by the public of loud fluctuating noises by removing the elements of uncertainty and shock.

Compaction of soils for engineering purposes

In construction work such as roads, runways or embankments, the controlled compaction of the fill material is of great importance. If before compaction the fill was closely examined, it would be seen that it was made up with soil or rock particles of various sizes, water and air voids. By compacting the fill the air is driven out and hence the density of the fill increased.

By increasing the density of the fill its shear strength is increased, its permeability and rate of water absorption decreased and the tendency of the fill to settle under repeated loading reduced. There is, however, a limit to the degree of compaction that can be achieved by mechanical means which will depend on:

The grading of the solid particles;
The moisture content of the fill, and
The method of compaction used.

A well graded fill medium is one that is formed with an even distribution of particle sizes. Figure 3.44 illustrates such a fill, and it will be noted that voids between the larger particles are filled by the smaller particles, producing a dense medium with minimum air voids. If, on the other hand, as Figure 3.44 shows, the fill is poorly graded, the size of the air voids is considerably increased and unless the compaction effort is sufficient to crush the particles, only a limited increase of density could be achieved.

The moisture content of the fill when being compacted is of major importance as there is a strong correlation between the achieved compacted density and the moisture content. In terms of soil mechanics the density achieved by compaction is known as the dry density. Figure 3.45 shows two curves on a dry density/moisture content plot. Curve 'A' is a theoretical curve representing the computed density or saturated dry density, of a fill medium where all air voids have been driven out but no water removed, for various values of moisture content. Curve 'B' represents the dry density achieved for various values of moisture content by

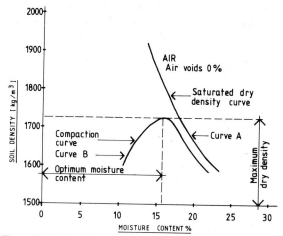

Figure 3.45 Dry density/moisture content plot

carrying out a suitable standard laboratory test. Curve 'B' is typical of that obtained by using the 2.5 kg hammer test method – BS 1377: *Methods of Testing Soils for Civil Engineering Purposes*. With a high moisture content the dry density achieved is fairly close to the theoretical value of the saturated dry density, although the dry density is low. As the moisture content is reduced the compaction achieved is increased until a point is reached where because of the reduced moisture content and hence lack of lubrication between the particles, the dry density achieved begins to decrease. There is, therefore, an optimum moisture content for a maximum dry density. This information and the fact that for any given moisture content value a related maximum dry density figure for the fill can be found is, of course, extremely important to the contractor. When laying the fill it is important not to attempt compacting deep layers as only the upper part of the layer may be fully compacted, producing an unstable fill of varying density.

Compacting plant

The majority of compacting plant operate on one or a combination of, the following methods of exerting compactive effort:

Static weight;
Kneading;
Vibration.

The decision on which type of compactor to use will depend on the physical characteristic of the fill medium. For example, clay – a cohesive soil, does not respond effectively to vibration as a compacting

Figure 3.46   The Aveling Barford general purpose compactor

Figure 3.47   The Aveling Barford self-propelled pneumatic tyred roller

Figure 3.48   Sheepsfoot roller

method, a combination of static weight and kneading proving the most efficient, whereas sand is ideally suited to compaction by vibration.

The main types of compacting plant are :

Smooth-wheeled roller – Figure 3.46
Pneumatic-tyred roller – Figure 3.47
Sheepsfoot roller – Figure 3.48
Vibratory roller – Figure 3.49

Smooth-wheeled rollers can be self propelled or towed by a track type tractor. The compactive effort is delivered through the self weight of the machine or the drum if towed. The most common form of this plant is the three wheeled smooth steel drum roller, popularly, but inaccurately, known as the 'steam roller'.

Pneumatic-tyred rollers are usually self propelled, the principal action is that of kneading. This is achieved with an odd number of smooth rubber tyres on two axles – the number of tyres used ranging from seven to nineteen. In order to achieve the greatest efficiency the pressure in the tyres should be adjusted to suit the fill and hence an air compressor must be in attendance.

Sheepsfoot rollers are hollow drum rollers with rows of protruding steel feet that knead the material as the roller passes over.

Vibrating rollers can be self propelled, towed or pushed. The rollers are normally smooth steel drums and the vibrations produced by a rotating eccentric shaft, for example in the drum. This type of roller is very effective in coarse grained soils such as gravel.

Table 3.4 summarises the various types of compactors, their general application and method of compactive effort.

Figure 3.49   Self-propelled vibrating roller

condition under the influence of specially designed poker vibrators. The action of the vibrator, usually accompanied by water jetting reduces the intergranular forces between the soil particles allowing them to move into a more compact configuration. After a certain compaction time, the particles are arranged in such a way that the optimum compaction has been reached. The vibrator is then raised and the procedure is repeated over the entire depth with backfill sand added simultaneously. As the soil becomes more dense during compacting operations, a crater forms around the vibrated area which confirms the effectiveness of the stabilisation process. The diameter of the vibrated area varies from 2 m to 4 m while the depth of compaction can be as much as 35 m below ground level – see Figure 3.50.

Vibro-compaction

A more specialist form of compaction is known as vibro-compaction which relies on the fact that particles of non-cohesive soil can be rearranged into a dense

Vibro-replacement

A further example of specialist compaction is the process of vibro-replacement. In this case a soil can be

Table 3.4 Summary of effect of compacting plant

Areas of application			Compactive effort
100% Clay .... 100% Sand .... Rock			
-----Sheepsfoot-----			Static Wt kneading
	Grid -------------		Static Wt semi kneading
---------------------Vibratory---------------------			Static Wt vibration
------------------Smooth steel drums--------------------			Static Wt
-------------------Multi-tyred pneumatic---------------------			Static Wt kneading

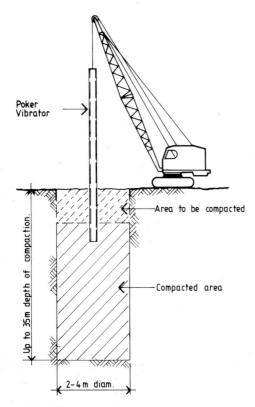

Figure 3.50   Vibro compaction

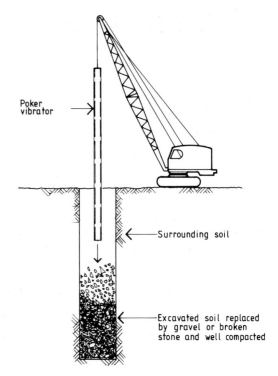

Figure 3.51   Vibro replacement

improved by partially replacing weak material with densely compacted granular columns tightly inter-locked with the surrounding soil. Normally, the columns fully penetrate the weak layer with the result that enhanced bearing capacity and settlement charac-teristics are achieved.

Because of their cohesive characteristics, clays, silts, and many layer soils cannot respond to the normal compaction methods and, therefore, this form of soil improvement can be used. The method involves the forming of a hole to the required depth with the specially designed vibrator. The hole is then filled with a coarse grained material consisting of gravel or broken stone. This fill is then compacted and a stone column is created. The bearing capacity of the stone column will depend on the quality of the fill material used, but standard tests will determine the ultimate bearing cap-acity of the fill.

Figure 3.51 shows the method of vibro-replacement.

4 Foundations

Influence of structural form, soil conditions and economics on choice

A structure, its foundation and the supporting soil interact with one another in a complex way. The behaviour of one depends upon, and influences, that of the others. Foundation design must, therefore, take into account the type of sub-soil supporting the structure, the materials used for the foundation and the economic feasibility of using a particular type of foundation. An approximate guide to the bearing capacity of some soils and rocks is given in Table 2.8.

Very often the type of foundation is dependent upon the form of construction used for the superstructure above. A small structure may only require a simple foundation, but as the height of a building increases so does the need for economy in the cost of foundations increase.

A foundation may be defined as that part of the sub-structure in direct contact with and transmitting loads to the ground.

Foundations may be broadly classified into two main groups:

(i) Shallow
(ii) Deep

Shallow foundations

This type of foundation is taken to be those where the depth below finished ground level is less than 3 m and include pad (isolated), strip or raft foundations. The depth to which shallow foundations should be carried depends upon three principal factors:

(a) Reaching an adequate bearing stratum;
(b) In the case of cohesive soils penetration below the zone where shrinkage and swelling due to seasonal changes are likely to cause appreciable movements;
(c) Penetration below the zone in which trouble may be expected from frost.

The selection of the appropriate type of shallow foundation will normally depend upon the magnitude and disposition of the structural loads and the bearing capacity of the soil.

Types of shallow foundations

Pad or isolated base (Figure 4.1)

This is a square or rectangular block of concrete carrying a single column. It may be reinforced or unreinforced depending on the column load and the bearing capacity of the soil. If reinforced the steel is placed near the bottom of the slab to resist the bending stresses set up by the double cantilever action.

This type of foundation is ideal when the columns are spaced far apart and a good bearing capacity is available at a reasonable depth.

Strip foundation (Figure 4.1)

This type of foundation may be used to carry load bearing walls. If the loads are light then an unreinforced concrete base will suffice, but if heavy loads are to be supported, it may be necessary to reinforce the base.

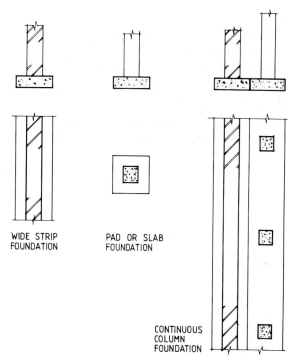

WIDE STRIP FOUNDATION

PAD OR SLAB FOUNDATION

CONTINUOUS COLUMN FOUNDATION

Figure 4.1 Foundation types (1)

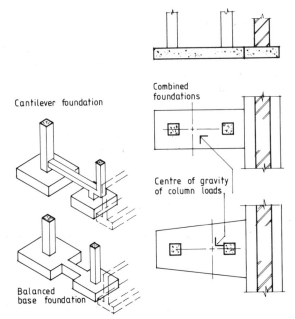

Cantilever foundation

Combined foundations

Centre of gravity of column loads

Balanced base foundation

Figure 4.2   Foundation types (2)

If the projection at the end of the base is restricted, then it may be necessary to use a trapezoidal base. This is necessary in order to permit the centre of gravity of the loads and centroid of the base to be in the same vertical line, since the position of the centroid of a trapezium along its axis may be made to vary with changes in the proportions of the ends.

Balanced foundation

This consists of the cantilever type which may be used as an alternative to a trapezoidal combined foundation or when some obstruction at a column position prevents an adequate foundation being placed directly under the column, for example when the column is placed close to the wall of an adjoining building.

Cantilever foundation (Figure 4.2)

The foundation consists of a ground beam, one end of which cantilevers beyond a base set a short distance in from the obstructed column and acting as a fulcrum to

Continuous column foundation (Figure 4.1)

This type supports a line of columns, and is one form of combined foundation. Circumstances when it may be used are:

(a) Where the spacing of the columns in one direction and the column loadings are such that adjacent pad foundations would be very close together or would overlap.
(b) Where there exists some restriction on the width of foundations at right angles to a line of columns.

This is common where a new building is to be erected between existing party-walls of adjoining buildings.

The strip is designed as a continuous beam, on top of which the columns exert downward point loads and on the underside of base the soil exerts a distributed upward pressure.

Combined column foundation (Figure 4.2)

In some cases where, for instance, a column base is restricted in its size by an obstruction, it may be advisable to link this column base to an adjacent base. In this way two columns (or sometimes three) are supported by the same base.

To ensure even pressure under the base, it is essential that the centre of gravity of the loads coincide with the centroid of the base.

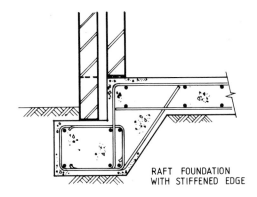

RAFT FOUNDATION
WITH STIFFENED EDGE

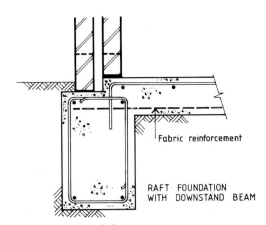

Fabric reinforcement

RAFT FOUNDATION
WITH DOWNSTAND BEAM

Figure 4.3   Raft foundations

the beam picking up the foot of the column, while the other end is tailed down by an internal column.

The counterbalancing force supplied by the internal column must be provided wholly by the dead load on this column. This must be at least 50% greater than the uplift due to the combined dead and live load on the outer column in order to provide an adequate margin of safety.

Balanced base (Figure 4.2)

It can be used when a base can be placed under, although eccentric to, the outer column and may be viewed as a beam balancing or resisting the tendency to rotate on the part of the eccentrically loaded foundation.

In design, the foundation base to the outer column is made large enough to take the column load. The tendency of this base to rotate due to the eccentric loading is resisted by a balancing beam linking it with the inner foundation base which, in turn, will tend to rotate under the action of the balancing beam.

The spread of this base must, therefore, be sufficient not only to distribute safely to the soil the load from its own column, but also to prevent overstressing of the soil at its far edge due to the rotational tendency.

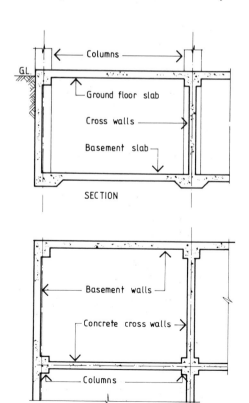

Figure 4.4 Boxed raft foundations

Raft foundation

A raft foundation is fundamentally a large combined slab foundation designed to cover the whole or a large part of the available site.

A raft may be used when the soil is weak and columns are spaced too close together or carry such high loads that isolated columns' foundations would overlap or would almost completely cover the site.

When a raft is indicated, the following considerations should be borne in mind before a final decision is made:

(a) When the depth of the weak strata down to firm soil is not much greater than about 4.5 m, it may be cheaper to use foundation piers.
(b) When the weak soil extends to a depth greater than this, piles might, in some circumstances, form a cheaper foundation than a raft.

A solid slab raft, see Figure 4.3 consists of a concrete slab reinforced in both directions. It is often economic only up to a thickness of 300 mm. If the slab is situated at ground level it is generally desirable to thicken the edge of the slab or to form a downstand beam of sufficient depth to prevent weathering away of the soil under the perimeter of the raft.

Deep foundations

These consist of foundations generally deeper than 3 m, but exclude piled foundations. The circumstances where this type of foundation is used would include the situation where no adequate bearing stratum exists shallow enough to permit the use of strip pad or raft foundations.

Careful consideration should be given to possible alternative forms of shallow foundations before deep foundations are considered. If circumstances dictate that a deep foundation is used, then advantage should be taken to increase the facilities of the project by the inclusion of, say, an underground car park.

Deep foundations may consist of basements or hollow boxes, brick or concrete piers and caissons.

Types of deep foundations

Basement or hollow boxes (Figure 4.4)

The main function of a basement is to provide additional space below ground level. It consists of concrete walls supported from the basement floor and is generally covered with a concrete slab at ground level. The loads from the superstructure are transferred through the perimeter basement walls and intermediate columns to the basement floor slab which acts as a raft.

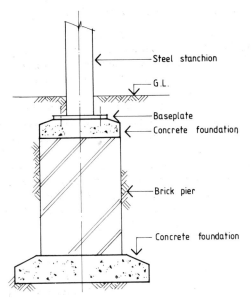

Figure 4.5   Brick pier foundation

Advantage may be taken of this form of construction to introduce reinforced concrete cross walls which are monolithic with the top and bottom slab and effectively divide the basement area into small compartments or boxes. This type of construction gives considerable rigidity to tall structures but they have an additional function in that they utilize the principle of buoyancy to reduce the net load on the soil. The buoyancy is achieved by providing the hollow box foundation at such a depth that the weight of the soil removed either balances the combined weight of the super- and sub-structure or is only a little less. The disadvantages of the hollow box type of foundation is that the small compartments are only suitable for plant rooms or stores but the true basement may be used for car parking or large storage areas.

Brick or concrete piers (Figure 4.5)
When a good bearing stratum exists up to 4.5 m below ground level, brick or concrete piers in excavated pits may be used.

In this case, a suitable foundation can be made by building up piers from low level to ground level and constructing independent bases to the columns or other supports on these piers at ground level. The piers are generally in the form shown in Figure 4.5 and can be constructed in brick or concrete. The maximum bearing pressure on top of the pier depends on the material the latter is constructed of.

Generally, it is better to provide as few piers as possible which means the piers are of generous proportions. Reinforced concrete columns can sometimes be

taken down to moderate depths but it may be necessary to provide lateral restraint at ground level.

Caissons
These are a form of pier foundation used when very heavy loads must be carried through water-logged or unstable soil down to a firm stratum. There are four main types of caisson:

(i)   Box caissons
(ii) Open caissons
(iii) Compressed air caissons
(iv) Monoliths

(i) Box caissons (Figure 4.6)

This type of caisson is usually constructed of concrete with vertical walls and bottom slab. The caisson is usually made in a fabrication yard and floated to the desired position, where it is sunk on to a previously well prepared base.

It is essential that precautions are taken against floatation. This can be achieved by flooding or ballasting the

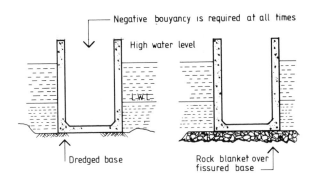

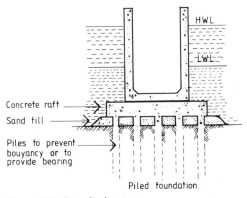

Figure 4.6   Boxed caisson

chamber or, if piles have been used to support the base, they will act as anchorage against buoyancy.

The void of the caisson is filled with concrete, using either a bottom opening skip pump or, if water is present, a tremie pipe. This type of caisson can be used as a foundation for bridge or similar supports.

(ii) Open caissons (Figure 4.7)

Similar in shape to the box caisson but are constructed in concrete without the bottom slab. The bottom of the walls are provided with a cutting edge to facilitate sinking through soft clay or silt. The excavation is carried out using conventional grabs which enables the caisson to sink under its own weight as work proceeds. This type of caisson is not suitable for sinking through ground containing large boulders. When the caisson has reached the required depth, the bottom is plugged with a layer of concrete to prevent water from entering the chamber. After pumping dry, the chamber is filled with concrete or ballast. Care should be taken to ensure that floatation does not take place before the chamber is finally filled.

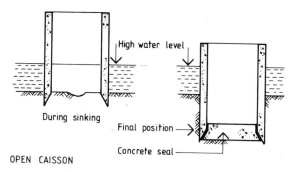

OPEN CAISSON

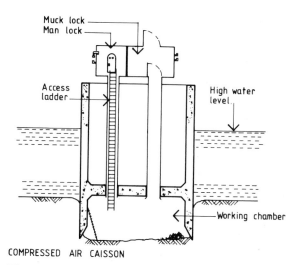

COMPRESSED AIR CAISSON

Figure 4.7 Compressed air caisson

(iii) Compressed air caissons (Figure 4.7)

This type is similar to the open caisson except that a working chamber is incorporated at the cutting edge. This chamber must be sufficiently pressurized to control the ingress of water or soil and to provide a safe working area.

The soil is excavated by hand and the spoil is removed by skip through a mud-lock. This method ensures that accurate levelling and testing of the subsoil can be carried out. The rate of sinking of the caisson is, of course, very slow. When the required depth has been reached, the working chamber is filled using successive layers of vibrated concrete and finally sealed with a pressure grout. The access shafts are sealed with concrete several days after filling the working chambers.

(iv) Monoliths

These are again similar to the open caissons except the walls are constructed of concrete of substantial thickness to resist any tendency to overturn. For this reason they are used for quay walls which have to resist impact forces.

This type of caisson is unsuitable for sinking through soft ground as it would be difficult to control the alignment of the structure.

Piled foundations

Introduction

The choice of a foundation for any particular structure will be influenced by many factors. Some of these will be economic, but the main factor will be the type and depth of the sub-soil encountered.

If a site investigation indicated that a firm stratum was available beneath deposits of soft clay or peat and this proved to be uneconomic to excavate through to provide a traditional foundation, then the decision may be made to use piles.

A piled foundation is used to transfer the load from a structural member (such as a column or wall) to a firm stratum, at some depth below the base of the structure, see Figure 4.8.

It is important to differentiate between the various conditions in which piled foundations are employed. Except in the case of larged bored piles over 600 mm in diameter, piles are seldom used singly.

Generally a group or cluster of piles are installed with a pile cap cast on the heads of the piles to distribute the load.

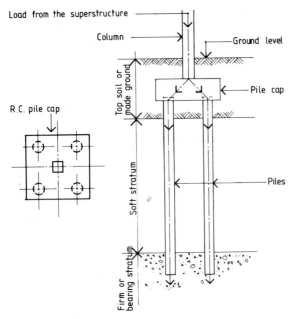

Figure 4.8 Typical fourpile cluster

End bearing and friction piles

The load from a superstructure may be transmitted to the surrounding strata either by; (a) end bearing, or (b) friction, see Figure 4.9.

In the case of (a), the piles are driven or formed through soft ground to derive most of their support by bearing on a harder stratum such as ballast lower down. Where no well defined hard layer can be reached at a reasonable depth, it may be necessary to use type (b) which rely for their support on the friction between their surfaces and the soil.

In practice, piles usually work on a combination of the two principles outlined above.

Pile classification

Piles are considered as displacement or replacement according to whether they are preformed and driven into the ground thus displacing the soil, or formed on site by drilling a hole and replacing with concrete and reinforcement.

Displacement piles are divided into three main types:

(i)   Totally preformed
(ii) Partially pre-formed
(iii) Driven cast-in-place

(i) Totally preformed

These preformed piles are manufactured from steel, timber or concrete and driven to the required depth.

(ii) Partially preformed

This type usually consists of concrete shells approximately 1 m long, connected together and driven to the required depth. Any surplus shells are then removed and the hollow shaft is filled with concrete and in some cases reinforcement is provided.

(iii) Driven cast-in-place

As an alternative to types (i) and (ii), the driven cast-in-place pile has been developed by piling contractors. In this case a steel tube with closed lower end is driven to the required depth and the tube filled with concrete. The tube may be withdrawn or retained in position after casting.

Replacement piles are divided into three main types:

(i)   Percussion bored
(ii) Rotary bored
(iii) Flush bored

(i) Percussion bored

A hole is formed by the repetitive use of a heavy cutting tool. When the required depth has been reached, a cage of reinforcement is lowered into the shaft formed which is then filled with concrete.

(ii) Rotary bored

With this method a hole is formed by extracting the soil using an open flight helical auger. The pile is formed in the same way as the percussion method.

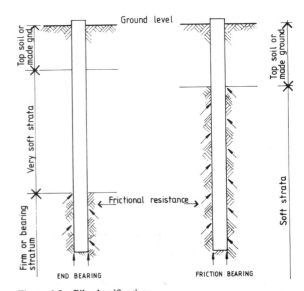

Figure 4.9 Pile classification

(iii) Flush bored

In this case, a hole, which is kept full of water or drilling mud, is formed by the use of a rotary tool. The water is continuously pumped out of the hole as the work proceeds thus removing the coarse and fine soil particles. On completion of the hole it is filled with concrete by using a tremie tube. This is necessary as the hole remains full of water.

In the case of (i) and (ii) it may be necessary to provide steel liners to the hole, particularly if the soil is loose or ground water is present.

Pile selection

The selection of the most appropriate pile type for a particular site depends upon many factors. These include type of sub-soil encountered (including groundwater conditions) size and form of the proposed structure, cost and location of site.

A proper choice of pile type can only be based on a thorough site investigation. Unless this investigation is carried out, serious difficulties may be encountered during the construction stage. It is therefore essential the following points are covered:

(a) The ground profile should be established to a depth below the base of the pile.
(b) Soil properties should be established by field and laboratory tests.
(c) Full record of groundwater levels.
(d) Full details of the boring operations which should include depth of bore and any casing required.
(e) The nature and method of boring through old foundations or other obstructions should be recorded.

The choice of pile type will depend on whether a displacement or replacement type is to be used. The main distinction between pile types is, therefore, made on the basis of soil displacement.

The structure and foundation should be considered together, since the selection could be influenced by the bearing capacity of the pile and could, for instance, lead to the adoption of large diameter piles instead of more numerous lightly loaded piles.

If the structure is in a marine environment, selection is always between concrete or steel displacement piles. The long-term durability and resistance to corrosion must be taken into account and the elastic properties of the materials should be considered. Bridge structures or viaducts in particular may employ piles extended above ground level to support the deck. The appearance of the pile could, therefore, be an influencing factor, as well as its ability to sustain bridge column stresses.

The ground conditions may have a profound influence on the selection of the pile type. For instance,

if the site investigation indicates that a cohesionless soil is present as the bearing stratum, then a displacement pile could prove to be the most suitable type. When a cohesive soil is predicted, a replacement pile should be considered. As mentioned previously, if water is present in the sub-soil and a replacement pile is specified, it may be necessary to line the hole as the work proceeds.

A comparison of the overall cost of one form of piling rather than another must form an important aspect of the pile selection. It may be necessary, for example, to compare the cost of large diameter piles against the use of a large number of small diameter piles. Other aspects may include the study of using an alternative foundation type. It may also be necessary to consider the additional cost involved if the site has a gradient of more than 1:20 and the installation of the piles requires the use of heavy plant.

The selection of piles could also be influenced by the location of the site.

Figure 4.10 Constructing piles in restricted area with limited headroom

The installation of displacement piles involves heavy driving which may cause considerable noise and, in some cases, excessive vibration. This could damage an existing structure in the close proximity of the site. In such cases a bored replacement pile may prove to be more suitable. When piling with limited headroom, piles of the percussion bored type should be used as the plant involved is light and easily portable, see Figure 4.10. If the site is in a remote area, restricted access may limit the type of plant which could be brought to the site and hence predetermine the method and type of pile used.

Use of various pile types

Precast concrete pile

A totally preformed concrete pile is suitable for open sites where the length of the pile required is constant. To be economic, this pile type should only be used on sites where at least 200 piles are to be installed.

Totally preformed concrete piles can also be used for marine works where conditions are likely to be suitable.

Table 4.1 shows the most common sizes for precast concrete piles and an indication of the maximum length available. Typical details of reinforcement required are shown in Figure 4.11 together with the lifting positions. It should be noted that the centres of the lateral reinforcement is increased at the head and toe of the pile to resist the stresses due to driving.

Table 4.1 Precast concrete pile details

Pile size (mm)	Maximum length (m)	Load range
250×250	12	
300×300	15	
350×350	18	Up to 1000 km
400×400	21	
450×450	25	

It is generally very difficult to predict the actual length of pile required, as sub-soil conditions vary across the site. This can result in the expense of cutting off and wasting excess lengths or of lengthening piles found to be short.

Precast concrete piles are driven by the use of a pile frame or driving rig, see Figure 4.15. A helmet is placed over the top of the pile, its main purpose being to hold the dolly and packing in place. This assembly is necessary to prevent the top of the pile splitting due to the action of the drop hammer. A variety of driving shoes are available and attached to the pile. The choice of shoe will depend on the ground conditions encountered.

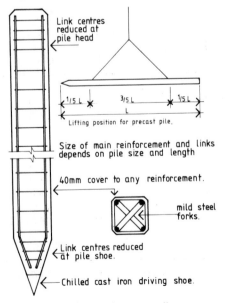

Figure 4.11   Precast reinforced concrete pile

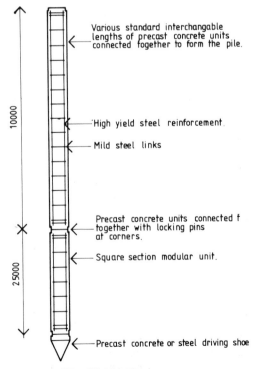

Figure 4.12   West 'Hardrive' system

Segmented concrete piles

To overcome some of the problems described above, the specialist piling contractors have developed a system based on short sectional precast units. As a result, standard lengths or modules of factory cast reinforced pile segments with integral coupling joints can now be driven and jointed quickly and easily. A typical example is the West 'Hardrive' system shown in Figure 4.12. Other proprietary types are available and Table 4.2 gives an indication of sizes and load capacity.

The segments are installed by the use of a pile frame or driving rig and linked together as previously described. A range of shoes are available according to ground conditions and a driving helmet is used to assist driving.

A further development has been the partially preformed pile. This generally consists of precast concrete hollow shells 1.0m in length, threaded over a special mandrel and driven to the required depth. Any surplus shells are removed and the shaft formed is then filled with concrete. In this way the problems associated with lengthening or shortening of piles are reduced.

Steel piles

This type of totally displacement pile is usually associated with marine structures. They generally consist of two types – Box Pile, which is manufactured from welded steel sections, or the more conventional 'H' section bearing pile.

Steel piles are installed by the use of a piling rig or frame as previously described, and the piles can be easily lengthened or shortened by welding and cutting. The high cost, and to some extent the effects of corrosion, could prohibit their use in certain circumstances.

Timber piles

Timber is rarely used for permanent piling work in the United Kingdom. Suitable timber for piling must be imported which greatly increases the cost of the material. There is also the risk that timber can be affected by rot or fungi. For these and other reasons, timber is now reserved for use in temporary works associated with civil engineering projects.

Driven cast-in-place piles

The use of this form of displacement pile should be considered when there is considerable variation in the depth of the bearing stratum thus resulting in different pile lengths. This type is also suitable where a weak strata overlies a firm stratum, or where a high water table is indicated.

Table 4.2 Sizes and loading capacities for various segmental piles

System	Joint type	Section shape	Segment size Width (mm)	Segment size Length (m)	Approximate axial load capacity (kN)
Balken Piling Ltd	Bayonet	Square	235 275	5 to 13	500 500 to 1000
GKN Keller Foundations Ltd	Spigot and socket with locking pins	Square	250 270 300	7, 10, 12	800 1000 1250
A Johnson Construction Company Ltd 'Herkules'		Hexagonal	[across faces] 220 305 340 390 480	4, 5, 6 7, 5, 9 12	700 1200 1300 1700 2650
'Herklid'	Bayonet	Square	250		500
West's Piling & Construction Company Ltd	Spigot and socket with locking pins	Square	285	5, 7.5	1200

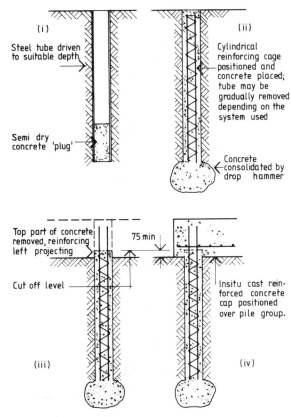

Figure 4.13 Sequence of installation of driven cast-in-place pile

The method employed is that a steel tube with a closed end is filled with a suitable semi-dry concrete mix (commonly known as a 'plug') to a depth of 1 m. The tube is then driven into the ground by a drop hammer operating aginst the plug, when a suitable depth is reached any surplus tubing is cut off. A cylindrical cage of reinforcement is lowered into the tube, and as the concrete is placed, the tube can either be removed or left in position depending on the method employed. The difficulty is to ensure that adequate cover to the reinforcement is maintained.

The disadvantage of this type of pile is that the plant used requires considerable headroom. Figure 4.13 shows the installation stages of a driven cast-in-place pile and its connection to the pile cap.

Rotary bored cast-in-place piles

Providing the site is reasonably level and ample headroom is available, it may be economical to use small rotary bored piles particularly if the sub-soil is of a cohesive nature.

An advantage of this type of pile is that installation is

virtually vibration free, and is, therefore, ideal when it is necessary to pile close to an existing structure.

The soil is extracted by the use of an open flight helical auger which, when rotated, fills the flights with spoil. When the flights are full, the auger is removed from the hole and the soil deposited, the operation is then repeated until a suitable depth is reached. Steel tubes are used if soil conditions demand to prevent the ingress of soil during boring. The tubes are removed after the reinforcing cage has been positioned and as the concrete is being poured.

An important development in foundation engineering has been the rise in popularity of the large rotary bored pile which can also have an under-reamed base. This type has proved to be suitable in supporting heavy column loads from multi-storey structures. The advantage being that in some cases one such pile can replace a cluster of small diameter piles, thus reducing costs, see Figure 4.14.

The installation of both types of rotary bored piles involves the use of heavy equipment which indicates they are not economic for small sites, and as Figure 4.15 shows, the equipment which can reach a height of 8 m or more, is only suitable for open sites.

A recent development in rotary bored piling has been the use of the high speed auger. By this method a 400 mm or 600 mm diameter pile can be bored using a hollow stem continuous flight auger of appropriate diameter and length. The walls of the hole are maintained by the continuous rotation of the auger. When the load bearing strata is reached, concrete is pumped under pressure through the rotary head and the hollow stem auger as it is withdrawn from the ground.

This method is particularly applicable when forming piles in water bearing sands, gravels and other unstable formations. During the boring operations very little soil is removed by the auger until grouting or concreting commences and, therefore, there is little danger of

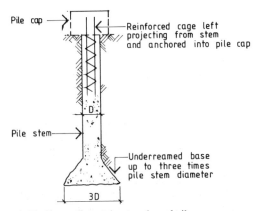

Figure 4.14 Large diameter rotary bored pile

disturbing the existing foundations of structures in close proximity to the piling works. No temporary lining tubes are required thus eliminating any noise and vibration.

Depending upon ground conditions, loads up to 700 kN can be carried by the 400 mm diameter pile and 1200 kN on a 600 mm diameter pile. It is suggested that a 20 m length of pile can be bored in 10 minutes.

The one disadvantage of this method is that a head-room of about 20 m is required for the piling equipment.

Percussion bored cast-in-place piles

This type of pile can be used in many situations such as sloping sites or piling in limited headroom, see Figure 4.10.

The method of construction is that a hole is formed by dropping a digging tool, which is usually a percussion cutter, when the tool is lifted it brings some of the soil with it. As the hole is formed, steel tubes made up in 1 m lengths and screwed together are driven into the hole. When a firm strata has been reached, a cylindrical cage of reinforcement is lowered into the hole, as the concrete is placed the steel tubes are gradually withdrawn.

This method is only suitable when the number of piles employed is small as the rate of boring is very slow.

Temporary liners

With all types of piling in which temporary liners are used, there is the problem that soil can squeeze into the hole once the liners are removed. This could effectively reduce the diameter of the pile. In the case of replacement piles, there is the additional difficulty in ensuring that the concrete cover to a reinforcing cage is maintained.

Bearing capacity and test loading

Bearing capacity of pile

The bearing capacity of a pile is dependent on the size, shape and type of pile and on the properties of the soil in which it is embedded. The ultimate bearing capacity is the load at which the resistance of the soil becomes fully mobilized. At a load greater than the ultimate bearing capacity the soil undergoes shear failure allowing the pile to penetrate into the ground, to become, in effect, a different pile from that originally installed; with a greater embedded length and possibly with different soil conditions.

The following methods may be used to calculate the approximate ultimate bearing capacity of a pile:

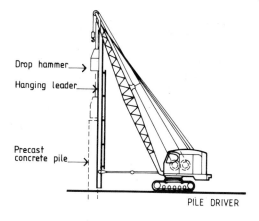

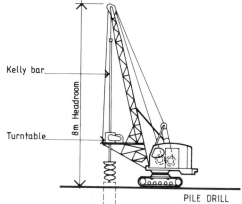

Figure 4.15 Pile driver and drill

(i)   Dynamic pile formulae
(ii) Static formula
(iii) Test loading

Briefly, by using (i), an estimate of the ultimate bearing capacity may be obtained from the driving characteristics of each pile, the accuracy being dependent on the reliability of the formula and data used. By using (ii) an estimated value of the ultimate bearing capacity of a typical pile is obtained; the accuracy being dependent on the reliability of the formula. The actual ultimate bearing capacity of any particular pile will differ from this value in so far as the soil surrounding it has properties different from those used in calculation. In the case of (iii), a test loading carried to failure gives the ultimate bearing capacity of the particular pile that has been tested. To find out the ultimate bearing capacity of other piles on the same site from loading tests above requires either the prior testing of an adequate number of piles and the use of statistical approach or a site investigation that is sufficiently detailed to show the uniformity or otherwise of the soil.

When driving or boring piles valuable information about site conditions can be obtained. Thus a judicious combination of load testing with driving and soil records can ensure reasonably satisfactory results on most sites.

Dynamic pile formulae

In cohesionless soils an approximate value of the ultimate bearing capacity of the pile may be determined by formulae. These formulae are based on the following assumptions, neither of which is fully justified:

(a) The resistance to driving of a pile can be determined from the kinetic energy of the driving hammer and the movement of the pile under a blow.
(b) The resistance to driving is equal to the ultimate bearing capacity for static loads.

These formulae are *not* directly applicable to deposits such as saturated silts, muds, clays and chalks.

If the use of driving formulae are restricted to those piles whose bearing capacities are at the toe in gravels, sands and other cohesionless soils of this type, then one of the more reliable formulae should be used which should give a calculated result within the range of 40% to 130% of the ultimate bearing capacity that would be determined by a test load.

There is no completely reliable formula but the **Hilary formula** is one of the more reliable and is the most commonly used in the United Kingdom.

$$R = \frac{Wh}{S + c/2} \quad n \text{ kN}$$

where c = the sum of the temporary elastic compressions in the pile, packings and ground (mm).
n = the efficiency of the blow of the hammer:

$$= \frac{(W + Pe^2)}{(W + P)}$$

where P = weight of the pile helmet (kN).
e = the coefficient of the restitution of the materials under impact.
W = the weight of the hammer (kN).
h = the height of the drop of the hammer (mm).
S = the final set of the pile per blow (mm).

The temporary compression of the pile and ground are found by actual measurements made while the pile is being driven.

Temporary compression of various forms of packings have now been established by laboratory tests.

Static formula

The weight of the pile and the load it carries are supported by the frictional resistance or adhesion of the soil on the surface of the shaft and the bearing resistance at the base of the pile.

The two resistances generally are added together and it is customary to assume that the ultimate bearing capacity of a pile is reached when both the resistances are fully mobilized.

This method has often been criticised for being too approximate compared with test loading and the dynamic formula.

However, one advantage is that the allowable loads can be assessed from the properties of the soils before work commences.

The normal tests for cohesionless soils include the standard penetration test and the Dutch or cone test (see BS 1377).

Test loading

In this case the test is made for the purpose of finding (a) the settlement to be expected at working load or multiple of, (b) determining the ultimate bearing capacity, (c) checking the structural soundness of the pile.

Pile tests are carried out using either of the following methods:

(i) Pile loading test using maintained loads;
(ii) Pile loading test at a constant rate of penetration.

In the case of (i) the test load can be applied in one of the following ways:
(a) Jacking against a 'kentledge' which is heavier than the required load – see Figure 4.16.
(b) Jacking against a reaction beam tied down by tension piles or ground anchors, these are located at a sufficient distance from the test pile to avoid influencing the test results, see Figure 4.17.

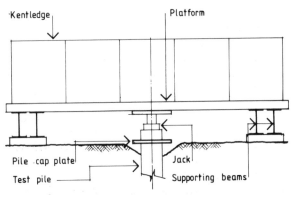

Figure 4.16   Pile test using Kentledge

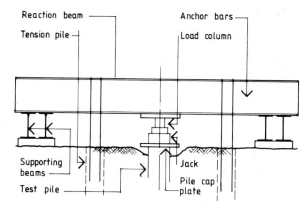

Figure 4.17 Pile test using tension piles and reaction beam

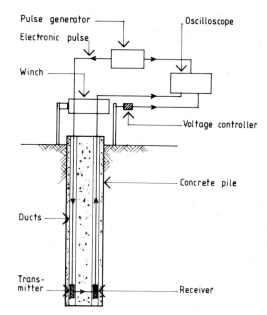

Figure 4.18 General arrangements for carrying out sonic coring

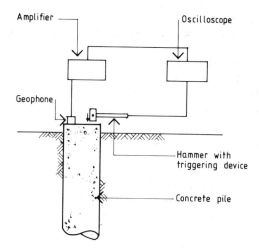

Figure 4.19 General arrangements for carrying out the seismic or echo test

The load is applied to the pile under test by an hydraulic jack seated on the head of the pile.

Settlements are measured either using a level and staff or by the use of general micrometer dial gauges fixed to datum frames situated well away from the test pile and loading system.

The procedure is for an incremental load to be applied in definite stages. At each level the load is held until the pile settlement has either ceased or has slowed to a specific rate of increase before the next load is applied.

A fairly recent development is the use of **integrity testing**. These methods are generally very rapid, cause minimal disruption to the site and are relatively inexpensive so that a large number of piles on a site can be economically examined. For example, where large diameter bored piles are used they support very heavy loads. It is, therefore, desirable to be able to confirm the structural soundness of every individual unit. On other sites, where small diameter piles are used, it may be only necessary to test a random sample of say 50%.

There are two main groups of test: simple integrity tests; and dynamic response methods.

Simple integrity tests

(a) **Sonic coring** – this method is based on measuring the propogation time of a sonic signal between two vertical tubes cast into the pile during construction. These tubes are filled with water to act as a coupling medium, two probes, one an emitter and the other a receiver, are lowered to the pile toe and raised in unison. As the probes are lifted a sonic profile is built up and recorded by a polaroid camera. Any defects are, therefore, clearly shown. A general arrangement is shown in Figure 4.18.

(b) **Seismic or echo test** – this technique is based on measuring the propogation time of longitudinal waves

in the concrete. The pile head is struck with a hammer and sends a compression wave down the shaft to the toe which is reflected back to the surface. A small device inside the hammer triggers a time base which together with other instruments measures the continuing length of a pile – see Figure 4.19.

(c) **Parallel seismic test** – this method is used as a 'last resort'. If the length or integrity of the pile is in doubt after the structure is complete, a small diameter hole is drilled adjacent to the pile and to its full depth. The hole is filled with water and a probe is lowered to the bottom of the tube and as it is raised a profile of signals is built up as hammer blows on the pile shaft send pulses down the pile.

Dynamic response methods

(a) **Vibration testing** – the pile head should be levelled with thermo-setting resin. The vibrator is placed centrally on the pile and a transducer fixed to a point close to the circumference. The test commences with the vibrator recording 20 Hz slowly increasing to 1000 Hz. The pile head velocity is recorded by a trace or pile signature. Considerable information regarding pile integrity and behaviour under load can be deduced from this trace.

(b) **Transient dynamic response** – this method is an easier way of obtaining the same results as recorded by vibration testing. The equipment is much lighter and consists of a small load cell, oscilloscope, geophone, microprocessor and plotter. The procedure is for the load cell to be struck by a hammer, this measures the magnitude of the dynamic force. The resulting velocity response contains the information needed to produce a V/F against frequency trace. This rather complex operation is carried out by a microprocessor and the resulting pile signature appears on the oscilloscope and is then copied by the plotter.

Pile caps

The primary function of a pile cap is to transmit the load from above onto the piles.

The two main considerations for selection of a suitable pile cap are: plan shape of cap and depth of cap.

(i) **Plan shape of pile cap** – the most compact arrangement of piles gives the most economic cap. However, the minimum spacing of piles permitted is controlled by the soil conditions and will normally lie between two and three times the pile diameter depending on whether they are end bearing or friction piles. An overhang of the cap beyond the outer piles is required for design and construction reasons which include sufficient room for the radius of bend of reinforcement and concrete cover, and the possibility of small inaccuracies in the pile positions.

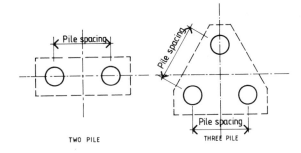
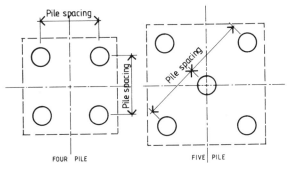

Figure 4.20 Various pile clusters

The suggested arrangement of pile clusters is shown in Figure 4.20 which determines the overall size of the pile cap.

(ii) **Depth of pile cap** – the depth is controlled either by bending or shear. In most cases, shear would be the critical factor in the cap design. Table 4.3 gives a guide to the overall depth of cap to the pile size. It is, therefore, desirable and economic to make the caps the same depth for the project.

Cap design

In practice pile caps are designed either by the trussed or beam method.

The beam method is the most popular, as caps can be designed for most pile clusters. Figure 4.21 shows the layout of reinforcement for each case.

Table 4.3 Relationship between pile size and depth of pile cap

Pile size (mm)	300	350	400	450	500	550	600	750
Depth of cap	700	800	900	1000	1100	1200	1400	1800

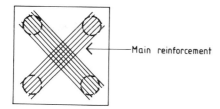

TRUSSED METHOD

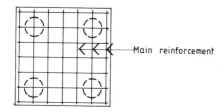

BEAM METHOD

Figure 4.21 Reinforcing pile caps

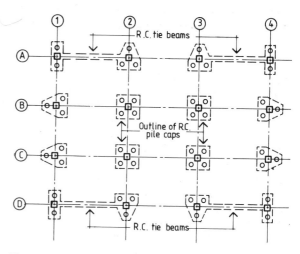

Figure 4.22 Piling layout

The piles should project into the pile cap a minimum of 75 mm and the reinforcement in the pile should be anchored securely in the cap to ensure a monolithic connection between the two.

Tie beams

Figure 4.22 shows a possible pile layout for a proposed structure. It should be noted that the two pile clusters are unstable about an axis, and should be tied back to a stable pile cluster by the use of a substantial reinforced concrete beam. In practice, it is usual to provide the beams between all the pile clusters. Their purpose is to provide support for the external cladding and in some cases the ground floor slab if it is designed as suspended.

Retaining walls

The function of a retaining wall is to hold, in a permanent vertical position, a wall of earth that in an unsupported state would be unstable. This condition will occur where a significant and abrupt change in ground level is required. Figure 4.23 shows the position of forces acting on a retaining wall.

The total active pressure due to retained soil is given by the equation:

$$P_a = K_a \gamma \frac{H^2}{2} \sec \delta \ (\text{kN/m})$$

D. Depth of the foundation foot below ground level at the front of the wall.
f1. Pressure on the foundation soil at the toe of the wall.
f2. Pressure on the foundation soil at the heel of the wall.
H. Vertical height of earth retained by the wall and the foundation.
Pa. Total active lateral thrust per unit length of wall, on wall back, due to earth alone.

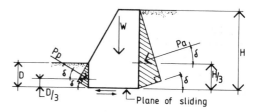

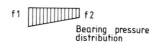

Bearing pressure distribution

Pp Total passive resistance of earth in front of wall, per unit length.
δ Angle of friction between retained earth and the wall back.
γ Average density of all strata down to a given depth.

Figure 4.23 Forces acting on a retaining wall

and the total passive pressure is given by the equation:

$$P_p = K_{p\gamma} \frac{D^2}{2} \ (\text{kN/m})$$

These equations assume horizontal ground with cohesionless soil backing where:

$$K_a = \frac{1 - \sin \phi}{1 + \sin \phi} = \tan^2 \left(45 - \frac{\phi}{2}\right)$$

and

$$K_p = \frac{1 + \sin \phi}{1 - \sin \phi} = \tan^2 \left(45 + \frac{\phi}{2} \right)$$

Typical values of ϕ, K_a and K_p are given in Tables 4.4, 4.5 and 4.6 respectively.

The stability of a retaining wall is dependent upon three main factors.

Overturning

A retaining wall will tend to overturn if the active pressure on the wall produces a stress at the toe of the wall which the material supporting the toe cannot resist; or the resultant thrust of the forces acting on the base of the wall is outside the base length, see Figure 4.24. It is a requirement of good practice to design the retaining wall so that there is a factor of safety of at least 1.5–2.0 against overturning.

Table 4.4 Typical values of ϕ for cohesionless materials

Material	ϕ (degrees)
Sandy gravel	35–35
Compact sand	35–40
Loose sand	30–35
Shale filling	30–35
Rock filling	30–35
Ashes or	35–45
broken brick	35–45

Table 4.5 Typical values for Ka for cohesionless materials for various valves of δ and φ (vertical walls and horizontal ground)

δ	φ				
	25°	30°	35°	40°	45°
0°	0.41	0.33	0.27	0.22	0.17
10°	0.37	0.31	0.25	0.20	0.16
20°	0.34	0.28	0.23	0.19	0.15
30°	–	0.26	0.21	0.17	0.14

Table 4.6 Typical values of Kp for cohesionless materials for various valves of δ and φ (vertical walls and horizontal ground)

δ	φ			
	25°	30°	35°	40°
0°	2.5	3.0	3.7	4.6
10°	3.1	4.0	4.8	6.5
20°	3.7	4.9	6.0	8.8
30°	–	5.8	7.3	11.4

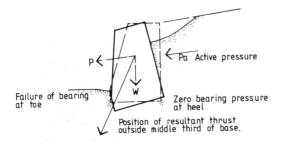

P← | ←Pa Active pressure

Failure of bearing at toe | W   Zero bearing pressure at heel

Position of resultant thrust outside middle third of base.

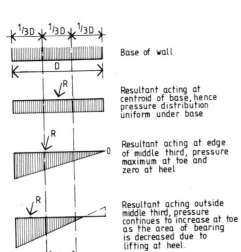

Base of wall

Resultant acting at centroid of base, hence pressure distribution uniform under base

Resultant acting at edge of middle third, pressure maximum at toe and zero at heel

Resultant acting outside middle third, pressure continues to increase at toe as the area of bearing is decreased due to lifting at heel.

Middle third.

Figure 4.24   Overturning of retaining wall

Sliding

Retaining walls will fail by sliding if the frictional resistance between the underside of the base and the ground on which it sits is insufficient to resist the active pressures, see Figure 4.25.

The frictional resistance can be increased against sliding if the passive pressure of the soil in front of the wall is brought into the calculation. In theory, the passive pressure will assist in resisting the forward movement of the wall due to active pressure. This should be treated with extreme caution as the passive

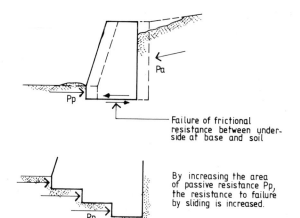

Failure of frictional resistance between underside at base and soil

By increasing the area of passive resistance Pp, the resistance to failure by sliding is increased.

Figure 4.25 Sliding of retaining wall

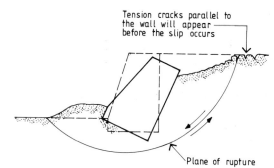

Tension cracks parallel to the wall will appear before the slip occurs

Plane of rupture

Figure 4.26 Circular slip of retaining wall

resistance of the ground is often sought at relatively shallow depths where the soil is subject to seasonal change.

In computing the total force resistance sliding, base friction or adhesion may be added to the passive resistance of the ground in front of the toe. A factor of safety of 1.5 is normally required against this form of failure.

Circular slip

Retaining walls supporting cohesive soils such as clay, may fail by a plane of rupture forming along a curve that will carry the wall forward and tilting it back as it goes. This is caused by the shear failures of the clay, see Figure 4.26.

Types of retaining walls

Retaining walls may be constructed of mass concrete, reinforced or prestressed concrete or steel sheet piling. The choice of material will, in most cases, be dictated by the form the wall takes. The various types of retaining walls may be classified by the various groups:

Gravity
Cantilever
Counterfort
Buttressed
Anchored

Gravity (Figure 4.27)

This type relies on the mass of the wall to resist overturning or sliding. These are usually constructed in either mass concrete and, to a lesser extent, brickwork.

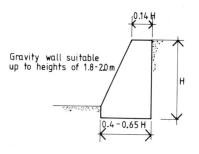

Gravity wall suitable up to heights of 1.8–2.0 m

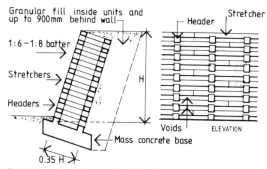

Granular fill inside units and up to 900 mm behind wall

1:6–1:8 batter

Stretchers

Headers

Voids

Mass concrete base

Header   Stretcher

ELEVATION

This type of wall is constructed of precast r.c. units built up to form a crib: suitable for embankments up to height of 5 m in single widths – as shown – or 9 m if double or triple depth headers are used.

Figure 4.27 Gravity retaining wall

Cantilever (Figure 4.28)

These rely on the strength of the stem at the base to resist the bending stresses induced by the pressure behind the wall. It must, of course, also resist overturning and sliding. Reinforced concrete is the most commonly used form of construction. The position of the base in relation to the stem will depend upon the location of the wall – that is, it is not always possible to

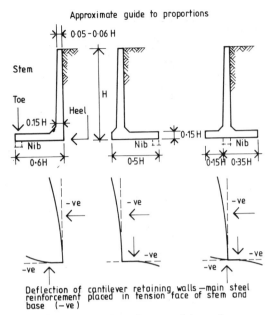

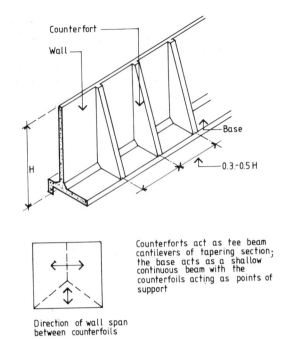

Figure 4.28 Typical RC cantilever retaining walls

Counterforts act as tee beam cantilevers of tapering section; the base acts as a shallow continuous beam with the counterfoils acting as points of support

Direction of wall span between counterfoils

Figure 4.29 Counterfort retaining wall

construct part of the base behind the wall although from a point of stability it is the position to be preferred.

Counterfort (Figure 4.29)

Although similar in cross section to the cantilever type, counterfort walls span horizontally between the counterforts apart from the bottom 1 m of the wall which cantilevers up from the base. The counterforts act as cantilever 'T' beams of tapering section and are usually spaced apart at distances of a half to one-third the height of the wall. This type of wall is always constructed in reinforced concrete.

Buttressed (Figure 4.30)

For high retaining walls (above 6 m) where it is not possible to excavate behind the wall, a buttress can be used. This type of wall is cast against the face of the excavated soil. Like the counterfort type the wall slab spans horizontally.

Anchored (Figure 4.31)

This type of wall is generally constructed using steel sheet piling. The horizontal support is provided by means of anchorages near the top, as well as by penetration of the sheet piling into the ground.

The principal types of anchorage used are mass concrete block, sheet piling, balanced above and below the

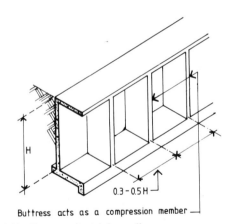

Figure 4.30 Buttressed retaining wall

waling, and sheet pile cantilever. They may consist of isolated concrete blocks or pile groups or form a continuous wall. The walings are usually made up from two steel channels back to back and fixed on the inside face of the wall. As the tie rod is a most important part of the structure it should be protected from corrosion by wrapping in bitumen based hessian or grouted in a non-ferrous casing.

Bored piles as retaining walls

As an economical alternative to sheet piling and, to some extent concrete cantilever retaining walls, contiguous bored piles can be used. The piles are generally installed using the rotary techniques which reduces noise and vibration which is useful if the wall is to be constructed in close proximity to an existing structure.

The choice of a piling rig depends upon site conditions and working room available. The reinforcement usually required in the pile shaft can be designed to resist heavy lateral loads in both temporary or permanent works.

Expansion and contraction joints

For continuous linear structures such as retaining walls, expansion and contraction joints must be provided for reasons of durability. In brickwork, vertical expansion joints should be provided at 5 m to 15 mm centres approximately. Reinforced concrete walls in the United Kingdom require vertical expansion joints at 20 m to 30 m centres and contraction joints to 5 m to 10 m centres. The reinforcement should not be continuous through the joints.

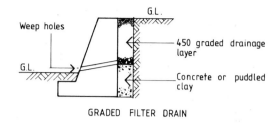

GRADED FILTER DRAIN

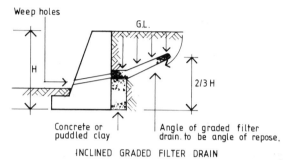

INCLINED GRADED FILTER DRAIN

Figure 4.32 Drainage behind retaining walls

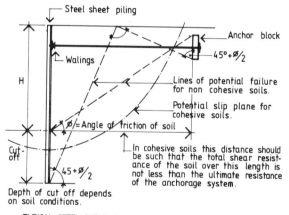

TYPICAL STEEL SHEET PILED, ANCHORED RETAINING WALL SHOWING APPROXIMATE POSITION OF ANCHORAGE IN COHESIVE AND NON COHESIVE SOILS

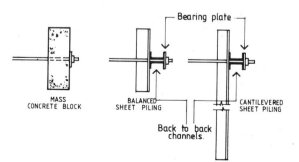

Figure 4.31 Anchorages

For convenience, pouring of the concrete should be carried out up to a contraction or expansion joints to ensure good keying on commencement of pouring the concrete the next day. If this is not possible a key should be formed by casting a bevelled groove or by exposing the course aggregate by hosing with high pressure water.

Backfilling

This can be defined as that portion of the material retained by the wall, which has been placed behind it after construction to fill the space between the wall and the natural ground. Adequate drainage of the earth behind retaining walls is of paramount importance since it reduces the surface water pressure on the wall.

The ideal filling material should be free draining and of good strength and durability. In general, suitable graded stone, slag, clinker or gravelly soils are the most satisfactory filling materials. The drainage system should consist of suitably spaced weepholes at least 75 mm in diameter. As an alternative, the water may be taken off by pipes at the bottom of the drainage layer and led to sumps or sewers. The drainage layer should consist of 450 mm of hand placed dry walling or large stones followed by a 200 mm filter layer of graded material. Figure 4.32 show details of drainage behind retaining walls.

Underpinning

The purpose of underpinning is to transfer the load carried on a foundation from its existing bearing level to a new level at a lower depth. This may be necessary for one, or a combination of, the following reasons:

(a) Settlement of foundations;
(b) To increase the load bearing capacity of the foundations;
(c) To allow works to be carried out below or adjacent to the foundations.

Before underpinning work is begun, the ground conditions must be investigated to identify the conditions responsible for any settlement. This will enable an appropriate system of underpinning to be chosen, also a general picture of the ground conditions and a measure of the bearing capacity of the soil on which the underpinning is to be supported.

It is very important that a full survey is prepared of the structure to be underpinned and any adjacent structures that may be affected by the work.

Records should be made of the levels of the floors and the inclination of the walls, marking, noting and photographing any cracks and/or defects of the structure. Glass 'telltale' is a strip of glass that is placed across a crack, and datum points should be placed where necessary for observing the movement of cracks and settlements. The report of the survey must be agreed by the owners of the properties, the contractor and engineers as a true record of the condition of the structures before work is commenced.

During the work a constant check must be kept on the structures by checking the original datum points and 'telltales'. It is also advisable, where possible, to remove as much 'live load' from the structure to minimise the load to be supported during underpinning operations.

Underpinning continuous strip foundation

The simplest form of underpinning to lower the level of an existing strip foundation is to carry the work out in a series of legs or pits. The length of each leg will depend on the spanning ability of the existing foundation. Generally for brickwork walls of normal type it is recommended that each leg should be about 1 m to 1.4 m in length but for walls capable of 'arching', a greater leg length may be used, see Figure 4.33.

Once the leg length has been excavated, the underside of the wall or foundation should be cleaned and levelled ready for pinning. The formation level of the leg should not be exposed until work is to commence, this requires that the last 100 mm or so of excavation be left until the levelling etc., of the old foundations has been completed. The leg should be constructed as

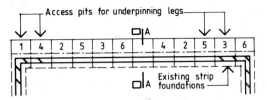

The maximum sum of unsupported lengths should not be greater than one quarter of the total length of the structure or one sixth of the total length if the wall shows signs of weakness or is heavily loaded.

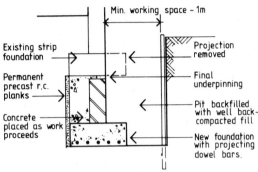

Figure 4.33　Underpinning strip foundation

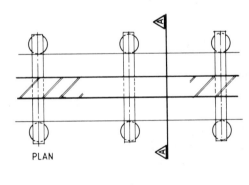

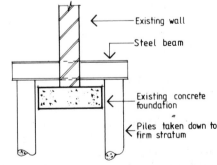

SECTION A-A

Figure 4.34　Underpinning strip foundation using piles

quickly as possible up to within 75 mm to 150 mm of the underside of the old foundation ready for final pinning.

When the leg has reached a sufficient strength to support the load to be placed upon it, the final underpinning can commence. This is carried out by ramming in a fairly dry concrete until hard up against the underside of the old foundation. The sides of the leg should be keyed to bond in with the adjacent legs and dowel bars placed in the ends of the new strip foundation.

The new series of legs should not be started until the preceding underpinning is completed and finally pinned.

If the depth of the formation level of the new foundation is too deep to use the above method, a piling arrangement may prove more economic. See Figures 4.34, 4.35 and 4.36.

These methods may involve some vibration, which would prove to be undesirable in many cases of underpinning. It is, therefore, obvious that the conditions of the structure and foundation must be taken into consideration before a decision can be made on the method to be adopted.

The most usual method is to install piles on each side of the wall connected at the top by a beam passing through the wall just above the foundation. The beam can be either reinforced concrete or steel, but steel beams are generally easier to fix in position.

As an alternative to the above, jacked piles, which consist of short pre-formed segments, are successfully forced into the ground using a powerful hydraulic jack operated by a pump. The segments are effectively connected together by inserting lengths of steel tubes into the segment hole which is then grouted in position, thus ensuring a monolithic connection. This method in installation almost entirely avoids noise, ground disturbance and vibration.

A different system has been developed which makes use of friction piles of various diameters. A hole is drilled at an angle, either directly through the concrete foundation or through the brickwork directly above to a pre-determined depth. The pile is completed by the introduction of a cage of reinforcement into the hole and the injection of cement sand grout pumped in under pressure.

Underpinning columns
In underpinning framed structures, the main problem is to provide a satisfactory support to the column which is being underpinned. Individual columns can be shored up by needle beams. In the case of steel stanchions, cleats are bolted or welded to the flanges to provide a support for the needles. For reinforced concrete columns, a chase is formed in the sides of the column and channel cleats are inserted and secured with trans-

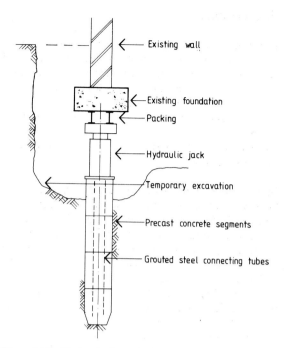

Figure 4.35 Underpinning strip foundation using jacked piles

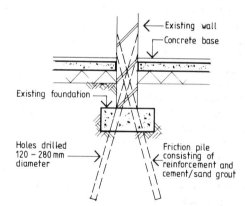

Figure 4.36 Underpinning strip foundation using friction piles

verse tie rods. A pair of steel joists act as needles and transfer the column load to temporary supports well away from the column, see Figures 4.37 and 4.38. Underpinning can then be carried out using either mass concrete or piling.

Underpinning adjacent to deep excavation
If a deep excavation is to be formed close to an existing structure, it may be necessary to underpin the wall adjacent to the excavation. It is often convenient to

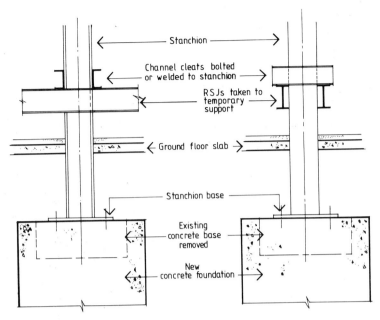

Figure 4.37   Underpinning steel stanchion

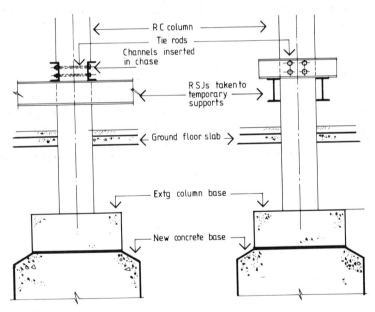

Figure 4.38   Underpinning RC column

combine the supports from the excavation with the underpinning of the existing structure.

For example, jacked piles can be installed under the existing foundation and contiguous bored piles can be installed which support the sides of the excavation (and in some cases act as a basement wall). Beams transfer the load from the existing foundation to the new piles, see Figure 4.39. It should be noted that the new piling should be designed to resist any lateral loads transmitted from the retained earth and in some cases from groundwater.

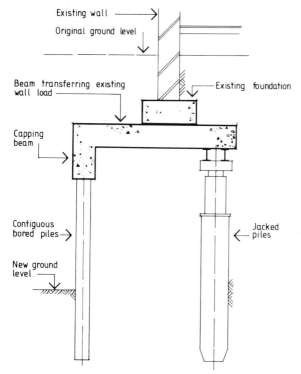

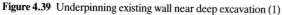

Figure 4.39 Underpinning existing wall near deep excavation (1)

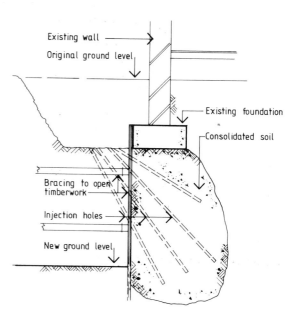

Figure 4.40 Underpinning existing wall near deep excavation (2)

Ground injections

Injections of the ground with cement grout or chemicals to fill voids or to strengthen the ground, are sometimes used as a means of underpinning.

Cement grouting is useful to fill voids in the ground beneath a foundation which has been caused by the erosion of loose granular soils. Chemicals can be used for injection into coarse sands or sandy gravels to form consolidated ground beneath a foundation and hence assist the underpinning process. In favourable conditions, chemicals can be used in connection with deep excavations close to an existing structure. The wall of consolidated ground acts as a retaining wall, thus reducing the need for shoring – see Figure 4.40.

5 Superstructure

The term superstructure when applied to construction is generally understood to mean that part of a structure above ground level, as opposed to the substructure which is below ground level with particular reference to structures designed as offices, hotels or residential use.

The traditional form of multi-storey superstructure is a framed construction composed of columns and connecting beams which support the floors and can be in insitu reinforced concrete, pre-cast concrete or structural steelwork. This system has the advantage of being simple in concept and construction, however it does have several disadvantages from a functional point of view. The main disadvantage is the presence of internal columns which tend to dictate the internal layout of rooms reducing the flexibility of use necessary in times when the change of use of a structure is not an unusual event, also the general external appearance can be somewhat basic although the thoughtful use of cladding can give a certain interest to the structure as a whole.

In order to increase the area of uninterrupted floor space the use of an internal reinforced concrete core has found considerable favour particularly with multi-storey construction. The core, constructed in most cases by slip-form, is used to house lifts, stairs and service shafts and occupies on average about 20% of the total floor area. One of several methods may be used to support the floors of structures built on this principle.

By cantilevering beams from the central core the floor areas are completely free of columns as is the perimeter of the structure which when clad with glass in the form of curtain walling gives an impressive sense of lightness to the structural form. However, the use of pure cantilever beams limits the size of the structures in terms of floor area on economic grounds and it is because of this fact that the use of propped cantilever beams is often adopted as an alternative form of support. The prop effect is obtained by using columns on the perimeter of the structure which can either be supported off a platform or beams cantilevered from the core at first floor level giving an unimpeded area at ground level or taken down to foundation level. In both cases the columns will be somewhat lighter than the equivalent external columns of a traditional framed structure. Another method is to suspend the external columns which are in this case usually of steel and act as ties, from large beams at roof level which are supported by the core. The floors in this type of construction span between the beams.

It is, of course, necessary to provide lateral restraint in most structures to resist wind pressure, and in medium and high rise construction this is achieved by either providing shear walls or bracing.

Shear walls can either be incorporated into the structure by the use of solid infill panels between the columns in traditional framed structures or by constructing the structure using solid walls in place of columns. A common arrangement is to use shear walls at two opposite ends with a core in the centre or two parallel walls running the length of the structure at right angles to the end walls to form a corridor. A method adopted by many local authorities for medium rise blocks of flats was to construct an egg box type structure of transverse walls in reinforced concrete or brickwork, the cells produced forming individual flats with partition walls to divide them into rooms. Access is provided by cantilevering the floor slabs beyond the face of the structure to provide walkways. An alternative method of resisting lateral wind stresses on traditionally framed structures is to brace the external columns using diagonal bracing. With hull core structures a rigid braced framework on the exterior of the structure called the hull acts with a central core to form a rigid structure.

The above descriptions cannot be considered in any way complete for students who are required to study in detail the construction of medium and high rise multi-storey structures, the subject being beyond the scope of this textbook and should, therefore, seek a more specialised text.

Structural steelwork

As a basic skeletal form for multi-storey framed structures, structural steel has several major advantages over the main alternative – insitu reinforced concrete.

The individual members such as the beams and columns are prepared (or fabricated) off-site and erected on-site, standard sections are used and the strength and consistency of the steel is strictly controlled under factory conditions. Problems inherent with insitu concrete work such as the fixing of the reinforcement, the placing and compacting of the concrete and the need for extensive falsework do not concern the steel erector. Only extreme weather conditions will stop the work, this is a major advantage in winter as steel erection can continue in low temperatures when concreting may have to be halted. The method in general can produce a considerable reduction in the on-site construction period of the frame leading to an overall reduction in the completion time of the contract.

From the point of view of quality control, on-site structural steel work has the advantage that all connections etc., can be truly inspected on completion where, as with insitu reinforced concrete, once the concrete has been poured, any disturbance of the reinforcement or, more seriously, omissions of reinforcement, are hidden from view and can only be checked with the aid of specialist equipment.

The initial stages in the design of a multi-storey framed structure is the same for reinforced concrete or steel. The architect on appointment will produce for the client general layouts and visualisations of the planned structure with approximate costs for the various schemes. On approval of a particular scheme the architect will appoint a consultant structural engineer to carry out the detailed analysis of the structure, the precise stage at which the structural consultant is called in will, of course, depend upon individual circumstances. At this point in the process, the stages for a steel structure differ to that of a reinforced concrete structure.

The consultant engineer on completing the main structual analysis of the frame prepares a series of drawings called framing plans. On these drawings are shown floor by floor the layout of the steel members noting the individual member sizes and the reactions to be resisted at each point of support. A further set of drawings showing the stanchion base layouts and column schedules will also be prepared. On completion of these drawings, a steelwork fabricator will be appointed either directly or by tender to either supply and erect the frame, in which case the fabricator will price the work on cost per tonne of steel to be erected or the main contractor may appoint the fabricator on a supply only basis, in which case the work will be priced on a cost per tonne of steel delivered to the site. It should be noted, however, that the two methods of placing a contractor described above are only typical examples – there are many variations in detail that the client through the architect or consultant engineer would enter into with a fabricator for the supply of the steel.

On receiving the contract, the fabricator's first task is to produce detailed designs of each connection based on the information given on the engineer's framing plans. The calculations and arrangement of the connections are checked by the consultant to ensure the arrangements comply with the initial assumptions by the consultants on the degree of fixity (ability to withstand a moment) of each joint used in their own frame analysis. On agreement, the fabricator will produce a materials list and place his order for the steel required. Depending upon the fabricator concerned, the steel may be supplied from his own stocks or from an outside stockholder. Shop details are also prepared at this stage. Let us now look in more detail at the framing plans, material lists and shop details.

Framing plans

Figure 5.1 shows part of a typical floor framing plan. It is essential that the individual members that together form the steelframe are clearly identified on the framing plan. The method adopted may vary according to the accepted practice of individual companies, but whatever method is used it must enable each member to be identified in such a way that not only the type, serial size and weight is clearly marked but also its location in the structure, to the most practiced eye there is little or no visual difference between a 305 × 165 × 40 kg/UB and a 305 × 165 × 57 kg/UB but with respect to the structural design, there certainly could be a difference.

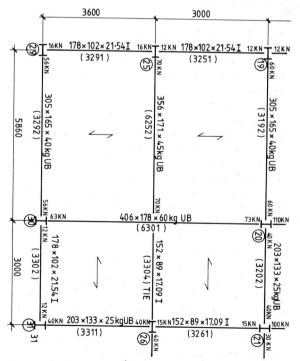

Figure 5.1 Part of typical floor framing plan

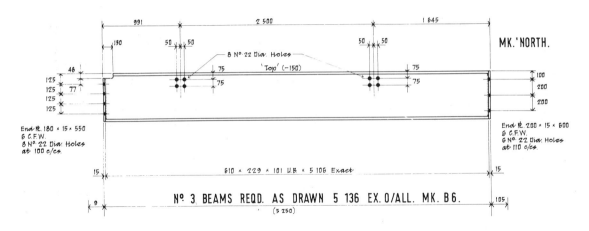

Figure 5.2 Typical shop detail of beam

The method of identification shown in Figure 5.1 is as follows. Firstly, each column is numbered and shown in its correct orientation according to the stanchion base layout. Each beam is represented by a single line which stops just short of its support points, this is important to avoid confusion where, for example, a beam supports another beam as in the case with beam 301 supporting beams 252 and 304. Along each beam is noted the serial size, mass and type, e.g.:

254×146		37 kg	UB
Serial Size	×	Mass per metre	Type of section

At each support is the vertical force to be resisted by the connection and in the brackets in individual identification mark. At first sight the mark might not seem to make much sense but is in fact designed to locate the beam on the framing plan according to the following criteria:

(a) The floor level.
(b) Whether the beam is shown on the framing plan as a vertical or horizontal line.
(c) Whether when looking at the line that represents the beam the left hand end is connected to:

    another beam, or
    a column.

(d) The column to which the left hand end of the beam is connected or, in the case of a beam to beam connection, the column to which the left hand end of the supporting beam is connected.

To identify the location of a particular beam on the framing plan is now relatively straightforward. Consider for example beam 3251. The first number 3 tells us that it is at level 3, and the numerals 25 and 1 that its left end is directly attached to column 25 and shown vertically on the framing plan. Referring to Figure 5.1, beam (3251) is a 178 × 102 × 21.54 kg I joist.

Shop details

As previously described, the framing plans and column schedules are prepared by the client's consultant engineer. On receipt, the fabricator will design the beam and column connections as previously noted, taking not only the engineer's design loads into account, but also any loadings due to the planned method of erection. The final connection details must be checked and approved by the consultant and on receiving approval the fabricator will prepare the shop details.

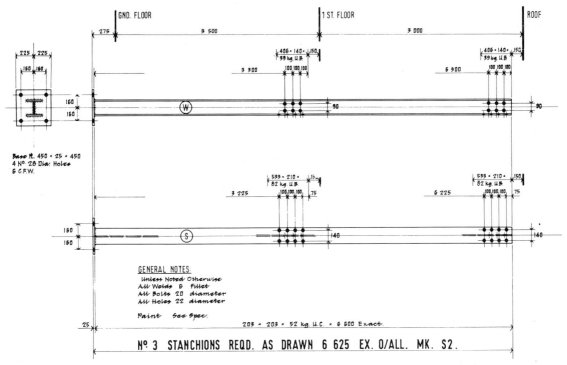

Figure 5.3   Typical shop detail of stanchion

Shop details are the drawings used in the fabricator's 'shop', that is the factory area where the steel work is prepared to produce or fabricate each member. Figures 5.2 and 5.3 show typical shop details of a beam and a column. For the sake of clarity it is usual practice to adopt two scales for the elevation of the member. The overall length of the member is drawn to a scale of, say, 1:50 in order to avoid using cut lines. But at a scale of 1:50, the depth of the members as shown on the drawing would be too small to show any detail clearly. A scale of, say, 1:20 or larger is used to represent the depth of the section, this allows sufficient room on the elevation to show holes and fittings although the centre line location of any fitting or hole is located by the 1:50 scale. Care in the dimensioning of the member is important not only in terms of overall dimensional accuracy, should a hole be drilled in an incorrect position the system of dimensioning should avoid the error being compounded by further dimensions being measured from the incorrectly positioned hole. One of two methods of dimensioning should avoid this problem; the first is locating each hole with a dimension that refers back to a single datum point for all dimensions, see Figure 5.2 or, alternatively, using what are called chain dimensions, an example of which is shown in Figure 5.3.

Steelwork connections

The joining of structural steel elements can be accomplished in one of several ways – that is with the use of structural steel fasteners (i.e bolts), welding or by a combination of bolting and welding. Whichever method is chosen it must satisfy a number of points – the two most important being:

(a) whether the method adopted will retain the structural integrity of the structure as a whole as envisaged by the structural engineers, and
(b) is the arrangement practical with respect to the requirements of shop fabrication and the on-site assembly?

The former point concerns the degree of fixity assumed to be present in the connection by the structural engineers for the purposes of the structural analysis of the completed structure and, although dealt with at some length in many publications, it still remains a topic of considerable discussion and research but is a topic beyond the scope of this particular work. The latter point of producing practical connection details is of course of interest to the student of construction technology. However, before considering the problem of producing practical arrangements, we must look in more detail at the structural fasteners and welding techniques available.

Structural fastener

The term 'structural fastener' applies in general to the range of bolts available for connections which have superseded in all but a few specialised cases the use of rivets. There are three main types of fastener available, each being fully described by the British Standards, they are:

(a) ISO metric black hexagonal bolts to BS4190
(b) ISO metric precision hexagonal bolts to BS3962, and
(c) High strength friction grip bolts to BS4395

ISO stands for International Standards Organisation.

Hexagonal bolts to BS4190 have a strength designation of 4.6 while hexagonal bolts to BS3962 have a strength designation of 8.8. The strength designation is interpreted as follows. The first number represents one-tenth of the minimum ultimate tensile strength expressed in the unit kgF/mm^2, while the second number is the ratio of the yield stress to the ultimate tensile strength × 10. Thus, for example, a bolt designated 8.8 will have an ultimate tensile strength of:

$$UTS = 8 \times 10 = 80 \text{ kgF/mm}^2$$

and a yield strength of:

$$YS = 8 \times 8 = 64 \text{ kgF/mm}^2$$

It has been the practice in the past to refer to bolts designated 4.6 as 'black bolts' – this, however, is not to be recommended, although bolts of 4.6 strength grade do in fact have a black finish, so do some 8.8 bolts; it is advisable, therefore, to distinguish between bolts to BS4190 and BS3962 by stating the strength grade, e.g. 6 M16 (4.6) or 6 M16 (8.8) on working drawings.

The term precision used with bolts produced in accordance to BS3962 refers not to the precision of their fixing or the size of hole into which they are positioned, but to the dimensional tolerance of the bolt shank. It is recommended in BS449 *The Use of Structural Steel* that both black hexagonal and precision hexagonal bolt receiving holes have a clearance of + 2 mm – that is, for example, a 18 mm diameter hole should be drilled for a 16 mm black or precision bolt. However, because the shank of a precision bolt is formed to a much closer tolerance, an allowance of + 0.15 mm to 0 mm may be specified if joint rigidity is

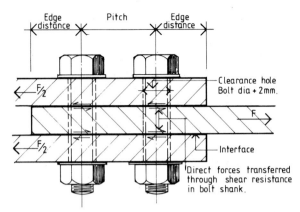

Figure 5.4   Transfer of direct force using 4.6 and 8.8 bolts

an important factor, it should also be noted that if a + 2 mm tolerance is supplied conditions such as fatigue or stress reversal except through wind loading cannot be accommodated. The mechanical principle of grades 4.6 and 8.8 hexagonal bolts is that the force is transferred from one element to another by either shear forces or tensile forces being resisted by the shank of the bolt, the applied force could also be a combination of both shear and tension. Figure 5.4 illustrates this point.

High strength friction grip bolts (HSFG)

Bolts to BS4395 are described in three parts according to their grade.

Part 1: *General Grade (strength grade 8.8)*
Part 2: *Higher Grade (strength grade 10.9)*
Part 3: *Higher Grade (waisted shank) (strength grade 10.9)*

Bolts to Parts 2 and 3 of BS4395 represent no more than 5% of HSFG bolts used in structural steelwork. It should also be noted that BS449, the general code for the design of structural steelwork, does not apply to the use of HSFG bolts. A separate code, BS4604 *The Use of High Strength Friction Bolts in Structural Steelwork* should be consulted when designing connections using HSFG bolts.

The HSFG bolt works, as its name suggests, by transferring forces through the frictional resistance of the contacting surfaces (called plies) of the connected elements. The frictional resistance is achieved by applying in effect a pre-stress force to the bolt shank which is therefore in tension. This force is in turn transferred to the contacting surfaces in the form of a compression force, the frictional resistance obtained by this clamping effect, between the contacting surfaces, is the basis

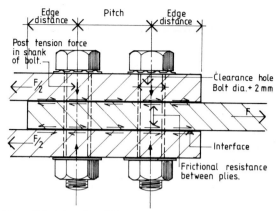

Figure 5.5 Transfer of direct force using HSFG bolts

of the strength of the connection. Figure 5.5 illustrates the principle.

The use of high strength friction grip bolts of grades 10.9 and 12 is restricted to connections subject only to shear. It is also recommended in BS4395 Part 3 that if a joint made with higher grade bolts carries the load in the direction of the bolt axis, waisted shank bolts should be used.

Methods of installing and tightening fasteners

Two conditions in general must be met for the satisfactory insertion of bolts into their receiving holes. The first condition is that the bolt thread and shank pass into the hole cleanly, that is the alignment of the hole through the members to be connected is good. Bolts must never be forced into position. Secondly, the surfaces to be connected should be at right angles to the axis of the bolt shank.

In the first case, if a bolt is forced into position because of misalignment of the receiving hole, the bolt may be damaged or at the very least jam during tightening. It is, however, the case that sometimes due to errors in fabrication, holes do not align, and then the holes must be reamed, that is drilled, in order to correct the mismatched. Care should be taken that the enlarging of the drilled hole does not reduce the edge distance to that recommended in BS449 Table 21.

The second point refers to the importance of the head of the bolt and the nut being tightened against a surface that is at right angles to the axis of the bolt shank. If for some reason this is not the case, a tapered washer can be used, this is particularly so with high strength friction grip bolts, where a tapered washer should be placed under the nut or bolt to be rotated and also under the non-rotated nut or bolt if the plane of the surface is outside the limits $90° +/- 3°$ to the axis of the bolt. All burrs around the edges of drilled holes

must be cleaned off and in the case of high strength friction grip bolts the surfaces in contact must be free of oil, dirt, loose rust, loose scales, paint or any other applied finish that may prevent the full frictional resistance allowed for in the design of the connection developing. If it is necessary to ease the threads of a high strength friction grip bolt with a high pressure lubricant to assist in tightening it must not be applied to the bolt until after it is inserted through its hole. In multi bolt connections, the bolts should always be tightened in a staggered pattern, and if there are more than four bolts in a group they should be tightened working from the centre of the group outward – see Figure 5.6.

Assembly and tightening of high strength friction grip bolts

The three methods of obtaining the corrected tension in a high strength friction grip bolt are:

(a) Part turn method
(b) Torque control method
(c) Direct tension indication

Part turn method

This method is only applicable to general grade bolts and is the most widely used. The bolt is assembled with a hardened steel washer between the head and the surface of the member or if the nut is to be turned a nut

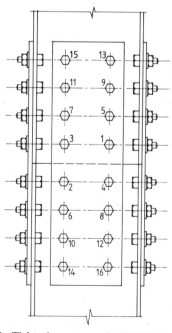

Figure 5.6 Tightening sequence for HSFG bolts

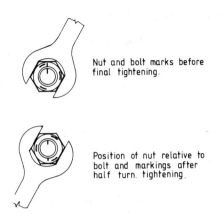

Nut and bolt marks before final tightening.

Position of nut relative to bolt and markings after half turn. tightening.

Bolt dia.	Bolt grip length / Rotation of nut Relative to shank	
	Not less than half turn	Not less than threequarters turn
M 16	Up to 150 mm	Over 115 to 225mm
M 20	Up to 150 "	Over 115 to 225 "
M 22	Up to 115 "	Over 115 to 225 "
M 24	Up to 160 "	Over 160 to 350 "
M 27	Up to 160 "	Over 160 to 350 "
M 30	Up to 160 "	Over 160 to 350 "
M 36	Up to 160 "	Over 160 to 350 "

Figure 5.7   Half-turn method of tightening HSFG bolts

face washer. Once the bolt is in position an opened ended spanner (podger spanner) is used to bring the surfaces of the connecting plies just into close contact. It is very important to ensure that the surfaces are in contact before the final tightening begins. Having obtained a close contact, a clear mark, see Figure 5.7, made with either paint or by using a cold chisel, is made on the nut and the protruding end of the bolt. Then according to the figures given in Figure 5.7, a half turn or three-quarter turn is applied to the nut relative to bolt using an impact wrench.

Torque control method

The torque control method is rarely used on site as more efficient methods are now available. The major disadvantage of this method is the need to calibrate the tool being used to apply the torque to the bolt. Such problems as the force required to overcome the frictional resistance between the threads, which will vary according to the condition of the threads, being included in the apparent direct tension being induced in the bolt requires careful calibration of the wrench which is both time consuming and requires specialist equipment. It has therefore been generally superseded by either the part turn method or by the use of load indicating bolts – see direct tension indication.

Direct tension indication

Of the three methods described, the method of direct tension indication, is used for the majority of installed high strength friction grip bolts. The basic principle of this method is that when the correct direct tension in the bolt has been reached, a visual indication is given. This is known as load indicating of which there are two main types. The GKN Lib bolt, see Figure 5.8, has four triangular pads which when initial tightening is complete are in contact with the surface of the member. As the bolt is finally tightened the pads yield and the minimum direct tension is achieved when the gap adjacent to the pads between the bolt head and the surface is reduced to 1 mm. The manufacturer however recommended that the gap should be further reduced to 0.75 mm to allow for a margin of safety. The gap is measured with a feeler gauge although it is suggested that experienced operators can judge the gap accurately by visual inspection. The second type of direct tension indicator available in the UK is the 'Coronet' load indicator – see Figure 5.9. With this method a special washer is used that has arched protrusions on one face. The washer is placed under the bolt head with the protrusions facing up so that there is a gap between the washer and the underside of the bolt head, a hardened steel washer is placed between the nut and the surface of the steel. When the bolt is tightened the protrusions depress and in a similar manner to that of the 'Lib' bolt, the minimum axial bolt tension is obtained when a specified width of gap has been achieved, normally this is in the region of 0.4 mm. If the nut is to be used for tightening then the load indicating washer should be placed under the nut on a nut face washer. It is not, however, necessary to place a hardened steel washer under the bolt head in this case.

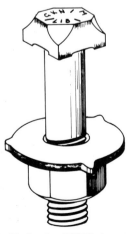

Figure 5.8   Load indicating (LIB) bolt

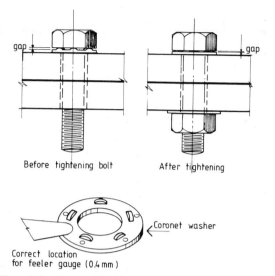

gap

gap

Before tightening bolt

After tightening

Coronet washer

Correct location
for feeler gauge (0.4 mm)

Figure 5.9 'Coronet' load indicator

When tightening has been completed, if for any reason the bolt assembly is slackened off, it must not be re-tightened but discarded and a new bolt assembly fitted.

Other high strength friction grip bolts

(a) Countersunk

If for aesthetic or practical reasons it is required that a bolt head should be as discreet as possible, a high strength friction grip bolt with a countersunk head is available in both general and higher grade strengths. The dimension of countersunk bolts are set out in BS4933.

(b) Close tolerance and turned barrel bolts

Suitable for connections where slip is not permitted. The use of these bolts has however in general purposes, been superseded by high strength friction grip bolts.

(c) Weathering steel bolts

Similar in both dimensional and mechanical properties to high strength friction grip bolts but available only in M20, M24 and M30 diameters weathering steel bolts are made from a special form of carbon weathering steel.

Recognition of structural fasteners

Bolts to BS4190 Grade 4.6

It is not normal practice for grade 4.6 bolts to be marked with the grade, the head has an M indicated on it and the manufacturer's trade mark.

Bolts to BS3892 Grade 8.8

It is mandatory for grade 8 bolts to be marked with an indented 8.8 and the letter M plus the manufacturer's trade mark.

Nuts to BS4190 and BS3692

The marking of nuts uses a code based on the face of a clock. At the 12 o'clock position a dot is indented and a figure 4 in the case of grade 4 nuts and an 8 in the case of grade 8 nuts positioned at their appropriate clock face locations.

Bolts and nuts to BS4395 Parts 1, 2 and 3

Standard high strength friction grip bolts to Part 1 of BS4395 are not marked with the grade but only an M (or ISOM) with three indents and the bolt again with only an M but three curved lines – see Figure 5.10. Bolts to Parts 2 and 3, that is grades 10.9 and 12 are marked with an M (or ISOM) and the grade as are the accompanying nuts – see Figure 5.11.

Ordering bolts

It is important when placing an order for bolts that the description of the required bolt type contains all the relevant information in a clear and concise manner, this

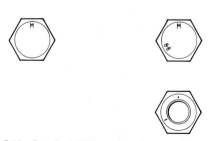

Figure 5.10 Standard HSFG bolt and nut markings

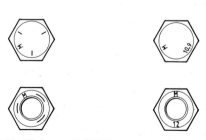

figure 5.11 Higher grade HSFG bolt and nut markings

will avoid the unnecessary waste of time and money that will be incurred if the order has to be re-issued. There is also the more serious danger of bolts of a lower grade or diameter than specified being used, the consequence of such an error being obvious.

It is therefore recommended that the following information be included in the description.

(a) The general product name such as 'precision hexagonal' or 'high strength friction grip'.
(b) The nominal diameter of the bolt preceded by the letter M to signify ISO metric.
(c) The nominal length, calculated as follows:
(d) Total length = bolt head + washer thickness + grip length (this is taken as the thickness of the connected plies) + nut + projection through nut (nominally 10 mm). Having obtained a length this is then rounded up to the next available nominal standard length of bolt.
(e) The appropriate British Standard number and part if applicable.
(f) The strength grade.
(g) If used, details of any protective coating including the appropriate British Standard number and part.
(h) A full description would read as follows:
(i) 'Precision hexagonal bolts M20 × 140 to BS3692'

Welding

As a technique for forming connections in structural steelwork welding has a considerable number of points in its favour. Essentially, a weld is a joint between two metal faces that has been formed by melting the metal which fuses with, in most cases, a suitable filler media to form a joint of at least equal strength to the parent metal. In comparison with bolted connections welded connections are simpler in their construction and have cleaner lines (an important factor with the present trend in architecture to 'expose' the steel skeleton of buildings) both points are graphically illustrated by comparing a bolted plate girder with an equivalent welded plate girder. Welded connections can, in certain cases, prove less prone to corrosion, for example vertical bolted joints will have ledges formed by the connected plates which tend to allow the accumulation of dirt and moisture which can penetrate the joint, the smoother lines of a welded plate connection will reduce the risk as the ledge will be sealed by the weld. Through welding the geometric forms available in steel members has increased considerably, structural members such as the castellated beam, tapered beam and open web girders are typical examples. The efficiency of welded joints in continuous structures such as portal frames and continuous beams has also led to lighter members being adopted.

Welding techniques

It should be appreciated by the reader that welding is a specialised and complex subject whose scope goes well beyond the brief introductory details outlined in this section. It is to be expected that such an important industrial process has generated numerous British Standards but, for the purposes of the non-specialist involved with welding, the following British Standards should at least be consulted:

BS499: *Welding Terms and Symbols*
Part 1: 1965: *Welding, Brazing and Thermal Cutting Glossary*
Part 2: 1980: *Specification for symbols for welding*
BS5135: 1974: *Metal-arc welding of carbon and carbon manganese steels*
BS638: *Arc welding plant equipment and accessories*
BS709: *Methods for testing fusion welded joints and weld metal in steel*
BS5289: *Code of Practice for visual inspection of fusion welded joints*

The techniques of welding available for joining metals are basically defined as either:
(a) Welding with pressure
(b) Fusion welding

The method used in the main for structural steelwork fabrication in the construction industry is metal-arc

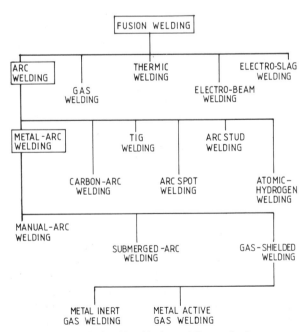

Figure 5.12   Relationship of fusion welding method

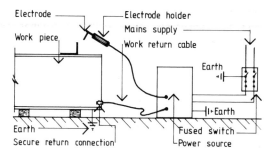

Figure 5.13 Typical welding circuit

welding which belongs to the fusion welding category. Figure 5.12 shows the relationship of the different fusion welding methods. It will be noted that there are three methods of metal-arc welding.

Metal-arc welding

The basic action of arc-welding is a controlled high temperature electric arc, 5000 to 30000 K°, that melts a defined area along the edges of the two sections to be joined causing the metals to fuse together. The current required to produce the arc, supplied by an appropriate AC or DC power source, is carried through an electrode and then arcs between the tip of the electrode and the area to be welded. The circuit is completed via an earth which is attached to the component being welded – see Figure 5.13. The electrode through which the current passes is usually a rod or tube of metal which melts at the tip where the arc has formed producing the weld pool, this particular type of electrode is called a consumable electrode. In the main, electrodes are coated with flux which is a material that melts on to the surfaces of the heat affected zone of the parent metals cleaning the surfaces and preventing atmospheric oxidation. The rate of disposition of the filler metal in the weld pool deposited by the melting of the electrode tip is controlled by varying the current and hence the temperature of the arc. The process is initiated by drawing the tip of the electrode along the surface of the metal and lifting it away to enable the arc to form.

Manual metal-arc welding (MMA)

As the name suggests this method of welding is performed by a welder and is therefore suited to either shop or site conditions. The electrode is held by the welder in a holder that also controls the current flow, see Figure 5.14. There are three types of electrode used, which one will depend on the type of weld and metal being welded. The three types are:

(a) Ductile: a general purpose electrode for steels with good welding properties. Ductile electrodes are in general used for structural steelwork fabrications.

(b) Basic: sometimes called a low hydrogen electrode it is used for welding steels with a high carbon content (0.4% and above) to reduce the risk of hydrogen cracking.

(c) Cellulosic: the particular advantage of this electrode is its ability to produce deep penetration welds.

It is important that electrodes are properly stored, they should be kept in the manufacturer's packaging until they are needed and kept in a dry and preferably heated store. It may be necessary in order to reduce the risk of hydrogen cracking to heat the electrode in an oven to drive off any moisture that may be present. Full details of this and associated procedures are given in Appendix E of BS5135.

Semi and automatic metal-arc welding

In principle, this method is the same as manual metal-arc welding, the difference being that apart from the element of automation involved, the electrode is fed from a drum on which it is wound.

Submerged-arc welding (SA)

Suitable only for shop welding as it is an automated system. A bare wire electrode is used with the flux added from a separate hopper where it is stored in powder form. The method produces a very consistent weld which is rapidly completed. However, as the flux is gravity fed into the arc it cannot work on areas with difficult access or on welds needing continued changes of direction, see Figure 5.15.

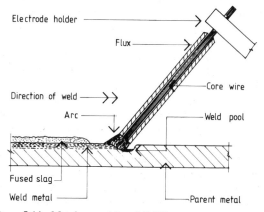

Figure 5.14 Metal-arc welding (MMA)

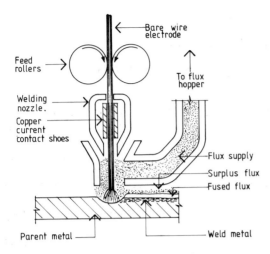

Figure 5.15   Submerged-arc welding (SA)

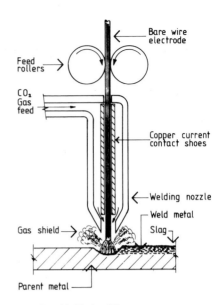

Figure 5.16   Gas shielded welding

Gas shielded welding

As with submerged-arc welding, gas shielded welding uses a bare electrode. A gas 'shield' is used in place of flux although certain welding processes are enhanced by the use of flux as well as the gas shield. For carbon and carbon manganese steels, carbon dioxide (CO_2) is used and this process is called metal active gas or MAG welding. If the metal to be welded is non-ferrous, then an inert gas such as oxygen is used. This process of gas shielded welding can be manual, semi-automatic or automatic, see Figure 5.16.

Safety precautions

As welding is a skilled and potentially hazardous process, it should only be undertaken by a fully trained welder, BS4871 and BS4872 describe the required testing procedures for welders. The arc produced during welding gives off two potentially dangerous forms of radiation, they are infra-red and ultra-violet. The effect of infra-red radiation on the human skin is commonly seen in the form of sunburn which can be both painful and leave scars on the affected skin in serious cases. It is necessary therefore to wear suitable protective clothing that covers all exposed areas of skin, this must include the area around the neck which is often exposed even when protective clothing is worn. In the case of ultra-violet radiation it is the welders' eyes that are at risk, the hazard being a complaint known as 'arc-eye'. Although arc-eye does not cause any permanent injury to the eye, it is nevertheless a very painful complaint which has been graphically likened to

having sand rubbed in the eyes and can last in severe cases up to twelve hours. In order to avoid arc-eye, a suitable head shield must be worn with special filter glass in the eye piece. The amount and type of shading used will depend upon the natural light conditions that the welder is working in and the type of arc being used. BS679 *Filters for use during welding and similar industrial processes* describes the appropriate filter shade to be used. It is also necessary to include in the eye pieces of the head shield a clear heat resisting glass to protect the eyes from the infra-red radiation.

The clothes worn must also give protection against splatter which is expelled as globules of molten metal. Persons working in the vicinity of welding may also need to wear eye protection to guard against the effects of ultra-violet radiation. Further details on safety procedures with particular reference to the plant and equipment used can be obtained in the Health and Safety at Work Booklet No. 38 *Electric Arc Welding*.

Types of weld

The terms joint and weld in the context of welded connections are often used to imply the same sense. This is incorrect. The term joint refers to the configuration of the parent metals to be joined and weld refers to the type of weld used to effect the joint.

The types of joints used in steelwork are numerous, the types of welds used are few. There are four basic types of weld, two of which account for 99% of welds made. They are butt and fillet welds, edge and fusion spot welds being the other two.

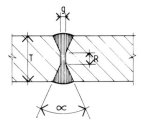

The full range of butt welds is shown in "Metric practice for structural steelwork".

Throat thickness of a butt weld depends on type of weld. Reference should be made to B.S. 449 (1969)

T = Thickness of plate
g = gap between plates
R = Root face
∝ Minimum angle

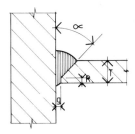

Refer to B.S 449 (1969) for welding definitions and strength of fillet and butt welds.
Refer to B.S. 449 (1980) part 2 for welding terms and symbols

Figure 5.17   Typical butt welds

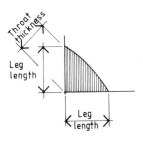

A fillet weld is specified by the leg length.
For 90° fillet welds, the throat thickness = 0.7 x leg length (or shorter leg length if unequal)

Figure 5.18   Fillet weld

The butt weld

The butt weld is used to form butt joints, T-joints and corner joints. Figure 5.17 illustrates the terms used to describe the integral parts of the butt weld and illustrates typical joints.

The fillet weld

A fillet weld is approximately triangular in section and is used to form T-joints, corner joints and lap joints amongst others, a typical example is shown in Figure 5.18. Fillet and butt welds can also be used to form a compound weld.

Other welding terms

Actual throat thickness

The dimension used for the purposes of design in the context of strength.

Bead.   A simple run of weld metal on a surface.
Run.   The metal melted or deposited during one passage of an electrode.
Tack weld.   A weld used to assist assembly or to maintain alignment of edges during welding.
Effective length.   The length of continuous weld of specified size.
Intermittent weld.   A series of welds at intervals along a joint.
Welding positions.   There are five fundamental weld positions – flat, horizontal-vertical, vertical and overhead.

Steelwork erection

The erection of steelwork, as has been previously described, is usually carried out by a contractor who specialises in such work. Before, however, the erection of the steelwork can commence the construction of the foundations must be completed by either the main contractor or a specialist foundation contractor according to however the contract has been arranged. The contractor concerned with the foundations will work to drawings supplied by the client's consultant engineer and by the steelwork contractors; whichever the case will depend upon the complexity of the foundations. With reference to the steelwork, the drawings will include information such as the layout of the stanchion bases, their dimensions, bolt positions and base plate and formation levels.

Generally one of two methods will be adopted for the stanchion base connections:

(i)   By use of holding down bolts
(ii) By use of a recess pocket in the foundation

(i) Holding down bolts

The responsibility for holding down bolts in so far as the foundation contractor is concerned is that they are set into the foundations vertically with the specified embedded length and that they are embedded in such a way as to allow sufficient lateral movement to facilitate the placing and erection of the stanchion.

The first point needs little further comment other than it is wise on the part of the contractor to agree with the steelwork contractor the required level of the top of the foundation and the level of the top of the holding

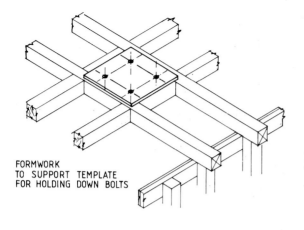

FORMWORK
TO SUPPORT TEMPLATE
FOR HOLDING DOWN BOLTS

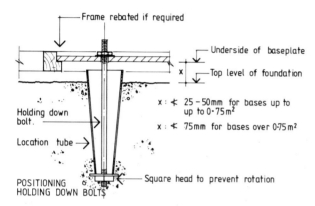

Frame rebated if required

Underside of baseplate

Top level of foundation

x : ≮ 25–50mm for bases up to
up to 0·75 m²

x : ≮ 75mm for bases over 0·75 m²

Holding down bolt.

Location tube

Square head to prevent rotation

POSITIONING
HOLDING DOWN BOLTS

Figure 5.19   Setting holding down bolts

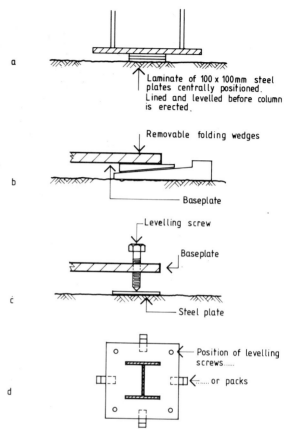

a

Laminate of 100 x 100mm steel
plates centrally positioned.
Lined and levelled before column
is erected.

Removable folding wedges

b

Baseplate

Levelling screw

Baseplate

c

Steel plate

d

Position of levelling
screws......

...... or packs

Figure 5.20   Adjustments for lining in and levelling stanchions

down bolts, that is how far the bolts should project above the foundation. It is considered good practice to allow not less than 50 mm between the top of the foundations and the underside of the base plate for base plates up to 0.75 m² in area and 75 mm are more for larger plates to allow sufficient room to grout in the holding down bolt pockets and the positioning of packs for levelling and lining the stanchions after erection.

The setting of the holding down bolts is not in itself a difficult task, however, unaligned bolts are a common fault. Two problems have to be considered when setting holding down bolts. The first is how to hold the bolts in position and at the correct level, the second, once having got them in the correct position and level is how to prevent them being displaced out of vertical during the pouring of the concrete.

To hold the bolts in their correct position the steelwork contractor may supply a steel template through which the bolts are passed, the whole assembly being supported at the required level off a frame. If such plates are not supplied the contractor can, of course, make up his own template. The holding down bolt

which will have a square block on the end to be embedded into the foundation to prevent rotation during the final tightening of the nuts is usually cast in a tapered wax cardboard, plastic or similar tube to allow lateral movement of the bolts. It is important that the tube projects above the finished level of the foundations to prevent concrete entering the tube during the pouring of the concrete. To prevent the bolts being pushed out of vertical alignment during the pouring of the concrete, the ends of the bolts should be secured for example by a couple of right angle pieces fixed to the reinforcement onto which the bolts can be fixed. Finally, the frame supporting the bolts if supported in turn on the side forms of the foundation may displace due to the side forms deflecting under the pressure of the wet concrete. This can be avoided by supporting the frame independently of the side forms. See Figure 5.19.

An alternative method which removes many of the problems outlined above is to cast the foundation and then using a tungsten carbide tipped rotary precision drill to drill out the pockets. For holes longer than 75 mm diameter a diamond tipped drill would be

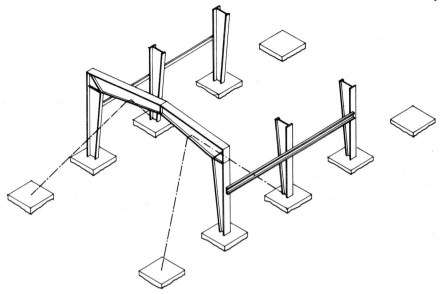

Figure 5.21 Erection of steel frame bracing (1)

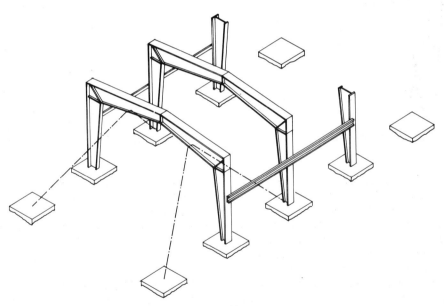

Figure 5.22 Erection of steel frame and bracing (2)

required. The holes must be thoroughly cleaned out using a combination of compressed air and water jet. Before pouring the concrete, however, it is necessary to make sure that the reinforcement is clear of the planned bolt positions.

(ii) Recessed pockets
For light to medium structures the use of a recessed pocket into which the stanchion is placed removes the need to use holding down bolts. The level of the base of

the pocket must be correct before the stanchion is erected as it cannot be adjusted once in place. If the base level needs adjusting steel or plastic packs approximately 100 mm square placed centrally and level in both directions can be used, and once levelled the packs should then be embedded in mortar. Timber wedges are placed between the pocket sides and the stanchion for lining in the vertical axis but must not be used to hold the stanchion erect – guys are used for this purpose.

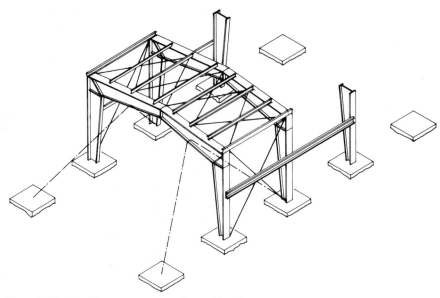

Figure 5.23   Installing eaves struts, purlins and bracing

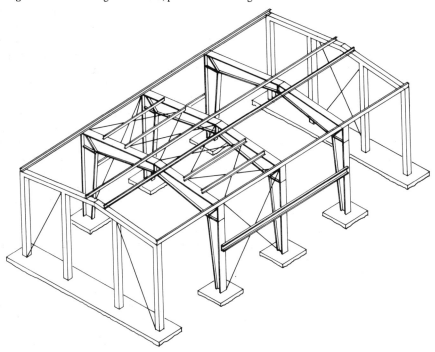

Figure 5.24   Erection of remaining frames

Erection of steelwork

Before the erection of the steelwork can begin, the steelwork contractor must check the position and level of the holding down bolts and the level of the top of the foundations. This work is often carried out when the contractor moves onto the site to start the erection work. This is not good practice for the obvious reason that if errors are discovered of a nature that require rectifying, this will lead to immediate delays. In order to avoid unnecessary delays of this kind, the steelwork contractor should check the foundation work on their completion thus allowing time for any remedial work if

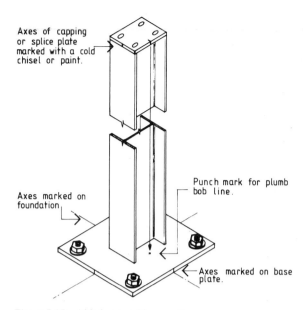

Axes of capping or splice plate marked with a cold chisel or paint.

Punch mark for plumb bob line.

Axes marked on foundation

Axes marked on base plate.

Figure 5.25 Aids for stanchion erection

required to be completed before the contractor moves onto the site in force to start the work of erection.

Apart from complete unalignment of the holding down bolts where it will be necessary to cut the bolts off below the surface of the foundation and set new bolts in drilled holes, minor errors can usually be accommodated. If the bolts are only slightly out of position, the receiving holes in the base plate can be slotted or the plate extended with new holes in the extension piece. The old practice of heating the bolts with a blow torch and bending them into position is not recommended. Bolts that are set too low, that is with insufficient length of thread for the bolts, can be dealt with using either a screwed ferrule or butt welding an extension piece on to the bolt. If the bolt projects too far above the stanchion base plate preventing the tightening down of the nuts washers used as packs or a sleeve can be placed between the nut and the base plate, however, it must be ensured that sufficient length of embedment is still available.

There are, of course, limits to this type of correction work and it is the responsibility of the client's representative on the site to rule on the acceptability or otherwise of suggested solutions put forward by the contractor concerned.

The first stage in erecting a stanchion having checked that the holding down bolt pockets are clear of any concrete, rubbish or water is to set erection packs on the foundations. Two types of pack are used. Thin steel or plastic plates which are built up to the correct level,

that is of the underside of the base plate, which will remain more often than not permanently in position, or removable folding wedges. An alternative method where very accurate levelling is required is the use of levelling screws. Typical arrangements of packs are shown in Figure 5.20.

Stability is the key word in the erection of steelwork. The contractor must adopt a scheme of work that will maintain at all times the stability of the incomplete structure. Stanchions when erected are normally held in position at first by a crane until guys are tied off to maintain stability, the holding down bolts must not be used for this purpose. Figures 5.21, 5.22, 5.23 and 5.24 illustrate a typical sequence of erection of a rigid portal frame structure.

To reduce the time required to line and level a stanchion a number of simple but effective preliminaries can be adopted, such as marking the axes of the stanchion on the concrete foundation and on the base plate, putting a punch mark on the base plate for the plumb line etc., as illustrated in Figure 5.25.

The erection of large steel framed structures requires a considerable amount of pre-planning. First the sequence of erection is decided upon. For a multistorey block type structure storey-by-storey (Figure 5.26) system is generally used. That is each storey is completely erected including the floor if it is also prefabricated before the next storey is started. The advantage of this system is that once the floor is in place work such as services, laying of floor surfaces etc., can start during the erection period of the steel frame. For structures that cover a relatively large area this method has the disadvantage that either a large number of cranes are required to cover the structure, or the cranes being used have to make time consuming lateral movements to cover the structure. In such cases the system of bay-by-bay (Figures 5.27 and 5.28) erection is considered more appropriate. Hence one crane is used to

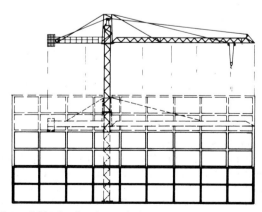

Figure 5.26 Steelwork erection 'storey by storey'

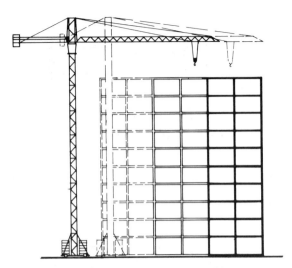

Figure 5.27   Steelwork erection 'bay by bay' (1)

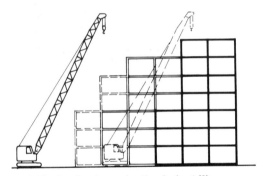

Figure 5.28   Steelwork erection 'bay by bay' (2)

erect one bay from ground to roof it will then re-position to erect the next bay.

To increase the rate of erection individual components may be pre-assembled on the ground and lifted into position. Typical components are vertical bracings, columns and main beams.

Having established the erection sequence the required lifting capacity, working radii and height of the cranes can be established. It is considered that in normal conditions, on an eight to ten hour shift one crane can perform between 20 to 40 lifting cycles – this includes slinging the component, moving it to its required location, unhooking and returning for the next component. In terms of tonnage of steel this means roughly between 10 to 15 tonnes per week. Together with past performance and the general experience of the steel erecting contractor a fairly accurate timetable

for erection can be built up which in turn will lead to a programme of delivery from the stockyard of the steel over a period of time. This is, of course, very important on constricted sites particularly in urban areas where storage is a problem.

Site connections, as previously discussed, are normally bolted. The first connection made between two members during erection will be temporary and where holes do not quite align due to lack of final lining a drift pin is used to ease the holes in line. Holes that do not align due to fabrication faults can either be slotted or redrilled. Under no circumstances should bolts be forced into position.

Typical steelwork connections are shown in Figures 5.29, 5.30 and 5.31.

Corrosion and fire protection of steelwork

The corrosion of steel may be defined as the gradual removal or weakening of the surface by chemical action which is always electrolytic. Rust, the yellowish brown coating formed on the surface of the steel is the result of the chemical reaction of the steel with oxygen and water and, in the absence of either oxygen or water, cannot form. The rate at which the corrosion will occur will depend on the exposure of the surface of the steel to water and oxygen but is accelerated by the presence of contaminates in the form of soluble salts, the two most common forms being chlorides and sulphates. Marine environments are particularly prone to chloride contaminates and heavy industrial areas, sulphates. A further source of sulphates in the form of sulphuric acid is acid rain. The effect, however, on the rate of corrosion of steel structures exposed to acid rain has yet to be fully examined.

To prevent the corrosion of steel that could in the most severe cases reduce the working life of a structure or, as is more often the case, cause expensive remedial work to be required the obvious action is to seal the surface of the steel from the main cause, that is water and oxygen. To successfully achieve this, however, care has to be taken in the preparation of the surface to produce a sound base for the protective coating. During the final coating of structural steel the surface reacts with oxygen to produce an oxide, blue grey in colour known as mill scale, over the entire surface. Although at first it appears as a sound surface the mill scale is in fact unstable and must be removed before the application of any coatings. However, it is not removed readily and only certain shop processes can remove it totally.

If the steel is to be prepared on site it is usual practice to allow the steel to weather, that is rust, to help loosen

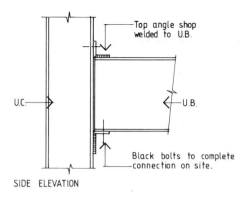

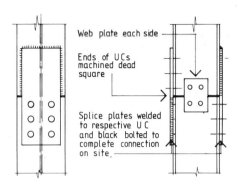

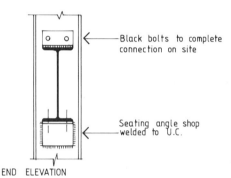

SIDE ELEVATION

END ELEVATION

Figure 5.29 Beam/column connection (welding and bolting)

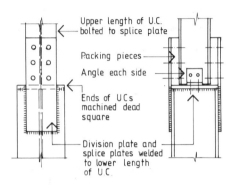

Figure 5.30 Column splices

the mill scale. It is then tackled either manually using scrapers and wire brushes or mechanically using, for example, rotary wire brushes. With either method no more than 30 to 35% of the mill scale and rust can be removed, it is recommended that red lead in an oil primer be used on surfaces cleaned by these methods.

Blast cleaning using either grit or shot blasting is used to remove mill scale and rust where a removal of not less than 80% is required. In this method abrasive particles are projected at high speed by either compressed air on a centrifugal impaler wheel. For site work expendable non-metallic grit is used but in shot blasting the grit or shot is metallic and can be re-cycled. BS4232 *Surface Finish of Blast Cleaned Steel for Painting* specifies three standards on the basis of the area of mill scale and rust removal. These are 1st quality – 100%, 2nd quality – not less than 95% and 3rd quality – not less than 80%. Shot blasting is preferred for most painting systems and grit blasting is essential for metal spraying.

Where steel is to be galvanised using the hot dip process acid pickling is used. This process requires the element to be submerged in a bath of acid which removes the mill scale and rust but does not appreciably attack the exposed steel surface and can be 100% effective.

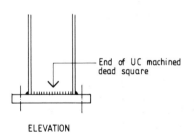

ELEVATION

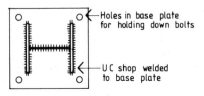

PLAN

Figure 5.31 Slab base

Protective coatings

The type of coating, the method of preparation and how it is to be applied will depend on a number of factors: the expected life span of the structure, the environment to which it will be exposed, the availability of shop or site treatment, the size and shape of the structural members and the cost being among the prime considerations. The coating may either be a paint based system or metal.

Paint systems are built up in layers, each layer or coat having a specific role. The first coat is the primer and has the two-fold function of providing a sound base for the subsequent coats and to act as a corrosion prohibitive. The intermediate coats are used to build up the thickness of the protective coating and finally the finishing coat which is the first line of defence against the environment. It also provides the finished look in terms of surface finish and colour. Table 5.1 shows the main generic types of paint and their properties. The column headed 'Solvent resistance' refers to the problem that the solvent in a paint may if applied to a previously coated surface affect that paint during application.

Metal coatings are usually either zinc or aluminium and can be applied in the case of zinc by hot dipping in a bath of molten zinc or spraying and by spraying for aluminium. Hot dipping is normally only used for items such as street lamps, crash barriers, balustrades etc., and not for structural elements. When applied by spraying the metal either in powder or wire form is fed through a special spray gun with a heat source, the molten globules of metal being blown onto the surface by compressed air. It is essential when using a sprayed metal coating to blast clean the surface prior to application.

The protection of structural steelwork by coating is not the only precaution available to the engineer, the detailing of the steelwork can also contribute. Good detailing should take into account the avoidance of unnecessary traps where dust and moisture can collect. Simple precautions such as positioning angles and channels face down are the provision of drainage holes

Table 5.1 Main generic types of paint and their properties

Generic type	Cost	Tolerance of poor surface preparation	Chemical resistance	Solvent resistance	Over-coatability after ageing	Other comments
Bituminous	Low	Good	Moderate	Poor	Good with coatings of the same type	Limited to black and dark colours. Thermoplastic.
Oil-based	Low	Good	Poor	Poor	Good	Cannot be overcoated with paints based on a sticky solvent.
Alkyel epoxy-ester etc.	Low–medium	Moderate	Poor	Poor–moderate	Good	Good decorative properties
Chlor-rubber	Medium	Poor	Good	Poor	Good	High build films remain soft and are susceptible to 'sticking'.
Vinyl	High	Poor	Good	Poor	Good	
Epoxy	Medium–high	Very poor	Very good	Good	Poor	Very susceptible to chalking in ultra-violet light.
Urethane	High	Very poor	Very good	Good	Poor	Better decorative properties than epoxies.
Inorganic silicate	High	Very poor	Moderate	Good	Moderate	May require special surface preparation.

Courtesy: British Steel Corporation

to allow water to escape and air to circulate can all help to prevent small areas of corrosion occurring. Access for cleaning should also be considered. For further information on this and other aspects of corrosion protection reference should first be made to BS5493 *Code of Practice for Protective Coatings of Iron and Steel Structures Against Corrosion*.

Weathering steel

As an alternative to protecting steel against corrosion by the application of coatings weathering steel may be used for certain applications and environments. The steel is produced under the trade names Cor-tan and Statevest. The particular characteristic of this steel is that it produces on its surfaces an initial rust layer that unlike that of ordinary steels becomes almost impermeable substantially reducing the rate of corrosion to a level that requires no further protection. However, in certain environments such as one with chloride contaminates it corrodes in a similar manner to that of mild steel, it also requires the use of special welding rods or bolts to obtain a uniform surface patina at connections. It is recommended that for the maximum benefit it should only be used in fully exposed positions in an industrial environment.

Fire protection of steelwork

The fire protection of structures and their individual members is required according to the Building Regulations if a structure has more than one storey, is built within a certain distance of the boundaries of the land on which the structure stands and has offices over more than one storey. The purpose of providing fire protection is first and foremost to ensure the safety of its occupants by controlling the spread of fire and maintaining the stability of the structure and its individual elements for certain specified minimum periods to allow the safe evacuation of the structure. The risk of the fire spreading to adjacent structures and the safety of those involved in fighting the fire are also important factors that contribute to the thinking behind fire regulations. The financial losses incurred both in terms of the fabric of the structure and the disruption of the operation of the company are also important factors. Insurance companies may often require fire protection of a standard beyond the minimum laid down by the Building Regulations by for example requiring the installation of sprinkler systems and smoke alarms etc.

Passive protection

Fire protection can be considered as either passive or active. The Building Regulations deal basically with protection of a passive nature, that is protecting structural members from the effect of fire and high temperatures for specified periods by the provision of coating or cladding the members. Active protection involves the introduction of systems such as automatic sprinklers which are activated by temperature or smoke sensors, this section deals only with passive protection.

The methods available for passive fire protection may conveniently be categorised under five headings:

(i)   Spray;
(ii) Boards;
(iii) Pre-formed casings;
(iv) Intumescent coatings;
(v)   Solid casing.

(i) Sprays

Sprays are normally based on either vermiculite plus a binder such as cement or mineral fibre and are applied wet. The period of fire protection afforded can be up to four hours depending on the thickness of the coat applied. Because of the unusual aspect of sprayed coatings they are, in the majority of cases, used on members that are concealed from view such as beams with suspended ceilings. The method is the cheapest and quickest of the five types listed but as it is applied by spraying, masking or shielding it may be required to control overspray, see Figure 5.32.

(ii) Boards

Boarded systems which are used generally to box the member, although on large sections it is recommended to follow the profile, are fixed either mechanically using, for example, screws or straps or by glueing and pinning. The board used ranges in thickness from 6 mm to 80 mm giving a maximum protection of four hours and particularly suited to vertical members such as columns. The board is made from materials such as

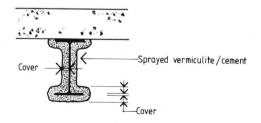

Fire resistance in hours	Min cover in mm
2	40
1	20

Figure 5.32   Profile protection for universal beams

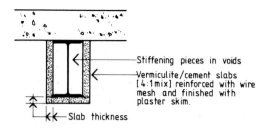

Stiffening pieces in voids

Vermiculite/cement slabs [4:1mix] reinforced with wire mesh and finished with plaster skim.

Slab thickness

Fire resistance in hours	Min thickness of slab in mm.
4	63
2	25
1	25

Figure 5.33   Hollow protection for universal beams

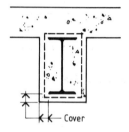

Cover

Fire resistance in hours	Minimum cover in mm.	
	Concrete contributing to beam strength	Concrete not contributing to beam strength.
4	75	63
2	50	25
1	50	25

Figure 5.34   Solid protection for universal beams

vermiculite or mica using cement or silicate in combinations or separately as a binder, see Figure 5.33.

(iii) Pre-formed casings

As the name suggests this type of system uses preformed sections which are produced in sizes to match the current range of universal beam and column sections. They are either 'U' or 'L' shaped and produced using reinforced plaster or sheet steel. Both boarded and pre-formed systems are within the medium to high cost range of fire protection systems but do offer the advantages of being applied dry and in certain cases providing a finished surface.

(iv) Intumescent coatings

Intumescence is defined as 'the forming, swelling and charring of a material when exposed to high temperature or flames'. The resulting effect of using materials that exhibit this characteristic as a coating is that a thick solid foam acting as an insulating layer is produced when exposed to fire on the surface of the steel giving up to two hours fire protection. The coating which is based on mastic or expoxy resins is used either as a thin layer for up to one hour protection applied by brush or spray or for two hours protection as a thick coat up to 13 mm, applied by trowel. This method is relatively new and at the time of writing is not approved for the fire protection of columns, its cost, depending on the time of fire protection required, can range from low to high.

(v) Solid casing

Solid casing using dense or light-weight concrete suitably reinforced is a traditional method of providing fire protection to structural steel members. The cost is high because of the additional requirements of formwork and the time to erect it. However, in certain circumstances, the concrete casting can be considered as contributing to the strength of the member, and this is particularly so in the case of columns, see Figure 5.34.

Construction of reinforced concrete and precast concrete suspended floors

Suspended floors

Suspended floors may be constructed in timber, reinforced concrete or precast concrete. The choice of floor type will depend on a number of factors. These include: load carrying capacity, fire resistance, sound insulation and economic span.

For the small structure the choice of floor is usually decided by the requirements of loading, span, cost, sound insulation, and speed of erection. In this case a timber floor may well prove to be suitable especially if the superimposed load is small. It should be noted that as the fire resistance is low and the sound insulation is less than a concrete floor the use of timber is usually limited to low rise domestic structures.

With a multi-storey structure other factors need to be considered when deciding on the type of floor to be used. The choice may depend on the type of structural frame, the degree of fire resistance to be provided and the sound insulation required. In this case an insitu reinforced concrete floor or precast concrete units may well prove to be the most suitable.

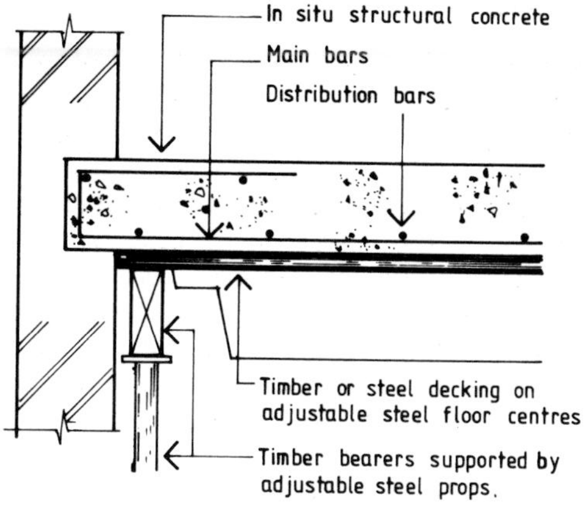

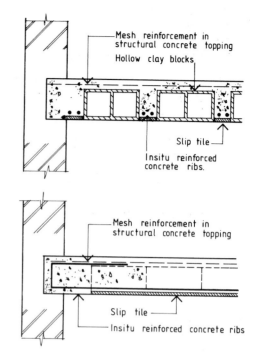

Figure 5.35 Suspended insitu reinforced concrete floor and formwork

Reinforced concrete floors

A simple insitu reinforced concrete floor slab consists of main reinforcement spanning one way between supports. Distribution bars laid at right angles and wired to the main bars are provided. The reinforcement used is generally of the high yield deformed type although a mesh reinforcement is sometimes found to be more economic.

Concrete has a greater fire resistance and better sound insulating properties than timber, although the deadweight of the concrete floor is much greater. Details of a reinforced concrete slab are shown in Figure 5.35.

The desirability of reducing the deadweight of the concrete floor to a minimum for reasons of economy has resulted in the use of the hollow block floor slab.

This type of floor is based on the flanged beam principle and consists of concrete ribs 75 to 150 mm wide and suitably reinforced. Clay or concrete hollow blocks are placed between the ribs and an insitu concrete topping must be provided, see Figure 5.36.

It should be noted that the hollow blocks contribute very little to the strength of the floor, but they do assist in providing a level soffit for plastering. In some cases a slip tile is provided under the rib to reduce the problem of pattern staining.

Precast concrete floors

Precast concrete floor units can be divided into five groups: hollow units, solid units, trough or T section, beam and block and concrete planks used as permanent shuttering. These units are illustrated in Figures 5.37, 5.38 and 5.39.

The first three types require a crane for handling, but no props are required during assembly. A small amount

Figure 5.36 Suspended hollow block floor slab

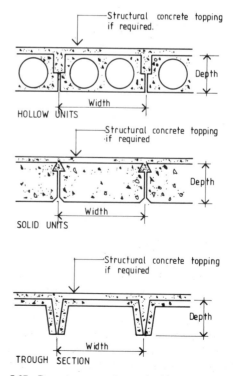

Figure 5.37 Precast concrete floor units (1)

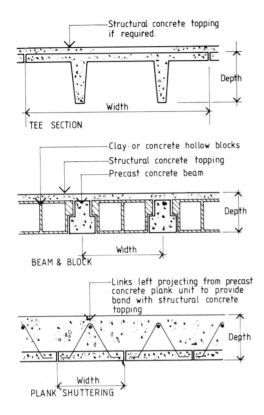

Figure 5.38 Precast concrete floor units (2)

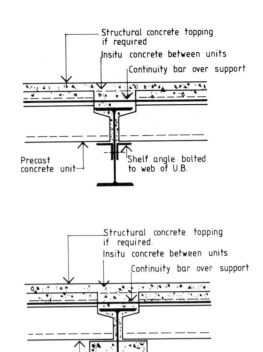

Figure 5.39 Details of supporting precast units on universal beam

of grout or concrete between the units is necessary and a structural concrete topping can be added to increase strength. A considerable advantage of these types of units is that they are able to carry their design load immediately after they have been erected.

The beam and block and concrete plank units require propping during construction and an insitu concrete topping is essential.

Floor units are an economic form of construction for a very wide range of spans. By using double-T sections it is possible to obtain spans in excess of 13.0 m. In the short span range precast concrete planks with an insitu concrete topping provide an economic decking. There is a very wide range of floor units that can be used in the intermediate span range.

The precast concrete floor is not suitable for irregular plan shapes which would require a large number of different sized units.

Some types of units are used in conjunction with a suspended ceiling in which case services can be placed within the voids. In other cases it may be necessary to accommodate the services within the screed or structural topping.

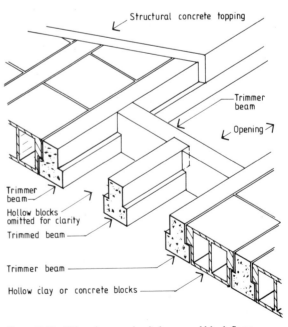

figure 5.40 Trimming opening in beam and block floor

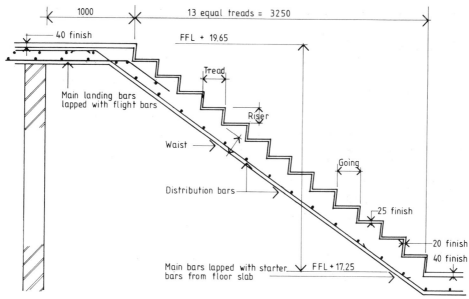

Figure 5.41 Section through concrete staircase

Trimming openings

Openings in precast concrete floors can be formed either by providing trimming beams or by using a mild steel strap. The method to be adopted depends on the type of floor used. Figure 5.40 shows a method of trimming an opening in a beam and block floor.

Openings in concrete floors can be trimmed by increasing the reinforcement in the thickness of the slab around the opening, but in some cases it may be necessary to provide trimming beams if the opening is very large.

Construction of a simple reinforced concrete stair

Reinforced concrete stairs

Concrete stairs are widely used because of their high degree of fire resistance and the many variations of plan layout that are possible.

Insitu concrete stairs may be constructed either by: cantilevering the flights from the supporting wall, spanning the flights between landings or beams, or the flight and landings acting as a single structural slab spanning between supports.

Where a crane is available, there may be some advantage in using precast concrete flights which span between insitu or precast concrete landings.

A wide variety of finishes can be applied to the treads and landings. The type will depend on the use of the stairs, but it should be noted that as the thickness of finish is greater on the floor than the treads, care must be taken when setting out the stair that the rise must be equal throughout.

Details of a simple reinforced concrete stair and landing are shown in Figure 5.41.

6 Bridges

Few structures are more representative of the work of the civil engineer than bridges and, although simple in purpose, that is the ability to bridge a gap, the structural design and construction can be one of the most challenging and complex aspects of modern civil engineering. With that thought in mind, a textbook of this nature can do no more in one chapter than introduce the reader to the basic principles of bridge works and deal only in general terms with the techniques of construction.

Bridge types

The classification of bridges by type can be approached from a number of ways. The traditional approach is to use the method of longitudinal support to the deck, hence the three descriptions of girders, arch or suspension are often applied. A girder bridge or beam bridge carries the deck on large girders which span on to piers or abutments. The girders can be steel or concrete, solid or lattice web. This form of construction, beams

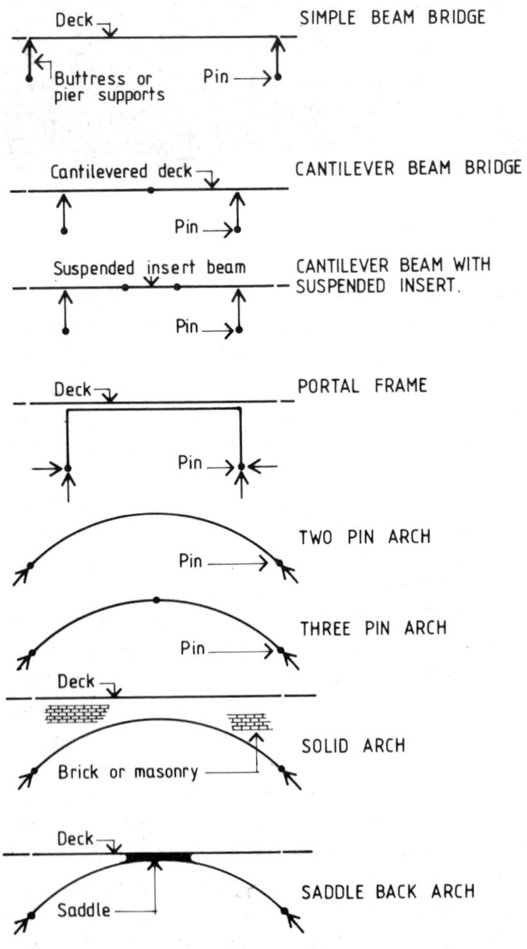

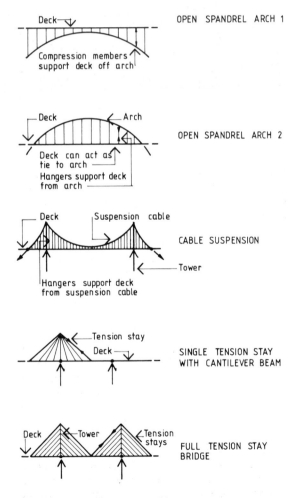

Figure 6.1 Basic bridge forms (1)

Figure 6.2 Basic bridge forms (2)

Figure 6.3 Gladesville bridge, Sydney, Australia

simply supported at their ends to form a single span, using prefabricated beams of steel or precast concrete are economic on spans of up to 40 m, and have been used extensively for minor bridge works. By placing the supports to the beams at a distance in from the ends of the beams a cantilever beam bridge is formed. The earliest form of this type of bridge structure was generally a symmetrical three span bridge of which each outer span was anchored at the ends and overhung into the central span by one third of its length. A suspended span was then placed on the cantilever ends to close the remaining third of the centre span. The Forth Railbridge built in 1890 is a famous example of this form of construction. Because the main supporting structure of this type of bridge is required to resist flexural stresses until the advent of steel and reinforced and prestressed concrete this form of construction had limited applications. Figures 6.1 and 6.2 show the basic form of girders and cantilever beam bridges.

Arch bridges have been used since Roman times in Europe. The reason being that the main load-carrying component of the bridge, the arch, is in a state of direct compression throughout its length, hence materials that are good in compression but poor in tension, such as masonry or brickwork, can be used to construct the arch. Figures 6.1 and 6.2 show some typical arch bridge components. Where the deck is below the crown of the arch as with through-deck arch bridges the deck can either be suspended by hangers from the arch ribs or incorporated into the structural form by acting as a tie to resist the horizontal component of the thrust in the arch.

Arch construction is less common today as a bridge form for a number of reasons. The high values of thrust involved in long span arch bridges require very substantial foundation works in comparison with other bridge forms of similar span. Unless the foundations can be positioned in rock or at least on a very hard material, their cost as a percentage of the total bridge works may be unacceptable. To prevent movement of the foundations it is often the practice to form a pinned connection between the arch rib and foundation, this is known as a two pinned arch. Another reason for the decline in arch bridges is fashion; like most structures bridges are to some extent affected by preferences of the day. In its favour the arch bridge does have the ability to create a high clearance at mid-span between the deck and the underlying water or ground, particularly in the case where the bank levels are low on either side, also because the arch can be skewed it is a useful form of construction to consider if the two banks are at different levels and the deck is inclined, see Figure 6.3.

For spans of over 250 m the choice of bridge type is generally accepted as being either suspension or cable-stayed. Suspension bridges have now reached spans of well over a kilometre, the Humber bridge having a clear centre span of 1410 m. The principle of this form of bridge is two continuous cables, which are anchored at their ends, carried over two towers forming a catenary between the towers. The deck is supported by suspending it from the cables on hangers. The towers which are cellular in construction can be built either in reinforced concrete using a slip forming technique or in steel using stiffened welded steel plate. Suspension bridges have not in general been regarded as suitable for heavy rail traffic and only one, the Salazar bridge in Lisbon has been designed for this purpose. The second form of cable bridge is the cable-stayed bridge which although not a new technique has in recent years gained considerably in use. In this case the deck is supported by tension cables which in turn are supported from a tower or pylon. The pylon can have one of several arrangements that is as a tower similar to that used on a suspension bridge where two sets of cables are used to support the deck on the outside on a single pylon either constructed as an 'A' frame or as a single tower. In the latter case of a single or 'A' frame pylon, one system of cables is used to support the deck along its centre line. The cables are arranged in most cases either radiating from a single point at the top of the pylon, this is called a fan (Figure 6.4) or from points up the pylon which is called a harp arrangement. Unlike suspension bridges the deck can either be reinforced concrete or steel as the tension stays impart a compressive force into the deck. This form of bridge construction has the advantage over similar span suspension bridges in that expensive anchorage blocks are not needed and an improved stiffness and aerodynamic stability is achieved, particularly if the deck is concrete. Under 150 m spans are not considered economic for cable-stayed bridges at present but in the 200–250 m span range they are seen as an economic alternative to segmental cantilever bridges. The longest span cable-stayed bridge constructed so far is the Yokahama Bay bridge in Japan with a centre span of 460 m.

Figure 6.4   Batman bridge, Tasmania, Australia

Bridge decks

Bridge decks can conveniently be divided into three groups on the basis of material used rather than structural considerations for the purposes of this chapter, they are concrete, composite (i.e. concrete and steel) and steel.

1 Concrete

The simplest form of concrete deck is a solid reinforced slab, however it is also the least efficient in terms of its span/weight ratio and is normally only used for short spans. When used it is often incorporated into a structure such as a box or portal to increase its efficiency by use of monolithic connections at the supports. The maximum economic depth is between 600 to 700mm as above this depth the self weight becomes excessive. To reduce the self weight voids are introduced into the slab to produce what is known as a voided slab, as shown in Figure 6.5. For practical reasons the voids are usually circular in cross-section and spaced with at least a 200 mm space between each void. The use of a circular void

with formers will assist the placement of the concrete, particularly beneath it. To form the void expanded polystyrene is generally used but care must be taken to secure the formers against flotation during casting. It is not a good idea to use the reinforcement as an anchorage as it has been known for the reinforcement to float up with the formers!

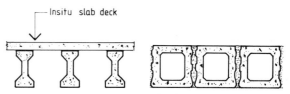

Figure 6.5   Reinforced concrete voided slab

Figure 6.6   PCDG standard section beams

Figure 6.7   Bridge spanning M53 mid-Wirral motorway

Insitu reinforced concrete voided slabs are considered economically viable up to spans of about 20 m to 25 m but above spans of 25 m prestressed voided slabs are generally used.

For short to medium spans pre-cast pre-stressed standard bridge beams have been in use for some time. They can be placed either in a beam and slab arrangement where an insitu structural topping is used or pre-cast slabs that span across to the beams or continuously. Figures 6.6 and 6.7 show some typical standard sections and arrangements. The advantage of this form of construction over the pre-stressed voided slab is the absence of falsework and the initial speed of erection although the advantage of cost will often remain with the latter method. Because of the problems and cost of transporting the pre-cast beams maximum span is limited to about 30 m. Longer spans can be achieved by casting the beams on site.

For longer spans single or multi-cell spine beams have been adopted in many cases, see Figure 6.8. This form of deck can be either pre-cast or cast insitu but is always post-tensioned. When pre-cast it is cast in short lengths which are erected using one of the techniques of segmental constructions. The sections can either form the complete cross-section of the deck or part of it; sections weighing up to 200 tonnes have been used. An unusual aspect of the pre-cast post-tensioned segment when constructed is the use of external cables. The pre-stress cables are suspended freely within the cells of the box, stressed in that position and subsequently protected by a casing of concrete. The main advantage of this system is the reduction of loss of pre-stress that occurs in ducted cables when tensioned.

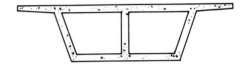

Figure 6.8   Twin cell spine beam

Figure 6.9   Composite steel and concrete bridge construction
Note: Shear connectors in top flange for concrete deck

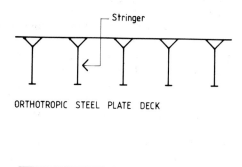

ORTHOTROPIC STEEL PLATE DECK

Figure 6.10 Deck profiles

2 Composite

A very common form of deck construction for short spans is the simply supported composite plate girder deck. The arrangement is similar to that of the concrete beam and slab deck except in this case welded plate girders are used with a reinforced concrete deck which acts compositely with the girders, see Figure 6.9.

3 Steel

For short spans if steel is to be used it will normally be in a composite form as described above. For medium and long spans a plate with longitudinal stiffeners or stringers is used. Figure 6.10 shows some typical cross-sections. Transverse ribs are generally used at spacing of about 1.8 m to 4.5 m and the deck plate which will be not less than 10 mm would have a finish of between 50 to 100 mm. Because of the particular problem of aerodynamics associated with very large suspension bridges the deck used on modern suspension bridges is an enclosed welded plate box section as shown in Figure 6.10.

Bridge abutments

An abutment to a bridge may, for the purposes of this chapter, be simply defined as the support to the end span at an embankment or cutting of the bridge deck. There are two types of abutment – wall abutments and open abutments, the latter also being known as bank seats. Wall abutments serve two purposes by supporting the deck and retaining the embankment or cutting whereas bank seats provide support only to the bridge deck.

Figure 6.11 shows typical wall abutments and bank seats with comments.

Piers

Piers like abutments also support the bridge deck but occur within the length of the deck rather than at the ends. The simplest form of pier to construct is a vertical member of uniform cross-section, although for aesthetic reasons piers are often of non-uniform section and have a textured surface.

If the deck of the bridge has a capacity to span transversely column type piers at suitable centres may be used but where no such capacity exists, such as in the case of box beam bridges, either solid reinforced concrete wall piers the full width of the deck or cross-head portal frames either in reinforced concrete or steel would normally be adopted. When single column type piers are founded on piles, although the required load capacity may suggest a single large diameter pile because of the problems that could arise due to inaccuracies in setting out the pile, it is considered more practical to use a group of smaller diameter piles. In order to allow the inspection of the reinforcement and any cleaning out that has to be done prior to pouring the concrete, piers should have a sufficiently large cross-section to allow a man to climb down inside if possible, this will also facilitate the placing of the concrete. Inclined or raked piers and 'V' piers are also often adopted, the latter in particular as the deck spans are reduced in comparison with vertical piers, and in both cases for aesthetic reasons. The problem with such

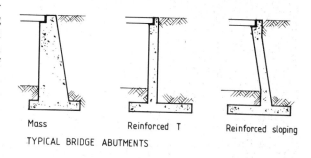

Mass   Reinforced T   Reinforced sloping

TYPICAL BRIDGE ABUTMENTS

Standard bank seat

Section   Elevation
Buried skeleton bank seat

Figure 6.11 Bridge abutments and bank seats

piers is the costly construction. Although not a particularly difficult problem in terms of technique, it is a costly process because of the greater complexity of the formwork and falsework plus the difficulty of obtaining a good surface finish free of blow holes caused by air trapped on the upper forms of the inclined members. Where raking piers transmit inclined loads through footings to the bearing strata it is, in terms of ease of construction, better to use horizontal and vertical bearing faces for the footing rather than attempt to construct a footing with its bearing face normal to the line of thrust.

It is the practice for certain types of bridge piers to insert a bearing to prevent horizontal forces and moments being transferred to the foundations of the pier. The bearing can either be at the top of the pier or at its base, although because of a number of disadvantages with the latter location, such as stability of the pier during construction of the deck, the top position is generally adopted.

Wing walls

Wing walls are used to either complete the visual aspects of a bridge abutment or to form an integral part, such as when the walls are parallel to the over-road. The walls usually constructed in reinforced concrete although crib type walls have been used in special cases, may be positioned parallel, angled or at right angles to the line of the abutment. Figure 6.12 shows some typical configurations.

Bridge bearings and expansion joints

All structures in varying degrees are affected by elastic deformation, creep caused by the material, in particular concrete, being subjected to a permanent force and thermal variation. In simple terms the effect on the structure is to produce internal stresses which, if not resisted, will cause movement such as, for example, thermal expansion or contraction. It is essential that such movement be catered for if the structure is to remain serviceable for its planned working life. So it is in the case of bridges provision must be made for movement, this is achieved by the use of bearings and expansion joints. Although both bearings and expansion joints are used to accommodate movement, bearings are also used to support the deck onto the piers or abutments whereas expansion joints are there purely to accommodate longitudinal movement of the deck due to thermal expansion and construction.

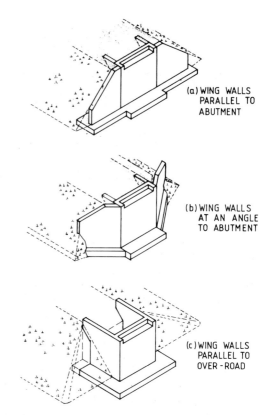

(a) WING WALLS PARALLEL TO ABUTMENT

(b) WING WALLS AT AN ANGLE TO ABUTMENT

(c) WING WALLS PARALLEL TO OVER-ROAD

Figure 6.12   Wing walls

Bearings

Up to three types of movement may be required to be accommodated by a bridge bearing, they are horizontal, longitudinal or transverse movement and rotational movement. For spans of less than 16 m provision is normally only required for horizontal movement, but for greater spans angular rotation of the deck at its points of support due to traffic loading, must also be accommodated.

The simplest form of bearing is a rubber strip or pad, this type of bearing is used to resist vertical loading and to a limited extent, rotational movement. Elastomeric bearings which are similar in performance to the simple rubber bearings but also permits horizontal movement by shear displacement, consist of steel plates laminated with vulcanised rubber.

When large horizontal or rotational movements are to be accommodated more complex bearings are adopted. For such bearings horizontal movement is accomplished by either sliding the contact surfaces of the plates being coated with a low friction material called polytetrafluoroethylene, or PTFE for short, which has now superseded the use of phosphor-bronze and, to some extent, rubber bearings. Where rotational

Figure 6.13 PSC 'sphericals' bearing for medium to heavy loads

movement is required in one direction only, usually about the transverse place, PTFE coated cyclindrical bearings are used and for multi-direction rotation spherically coated bearings. In practice, the bearing may well use a combination of the above such as a cylindrical rocker bearing activity with a second plane sliding surface. Figure 6.13 shows some typical bridge bearings.

The design life of bridge bearings, particularly the rubber and elastomeric types, is generally less than that of the bridge, bridges having a design life of up to 200 years. It is necessary when considering the position of the bearings to make due allowance for inspection and maintenance to prevent dirt and corrosion reducing the movement capacity of the bearing and if it is considered practical within economic constraints, to make provision for their removal and replacement.

Expansion joints

Expansion joints in bridge decks are used principally to accommodate horizontal expansion and contraction of the deck, this in turn requires provision to be made for the accompanying movement of the deck parapets, crash barriers, kerbs etc. The number of joints required will depend on the predicted maximum movement to be accommodated and the capacity of the joints to take up that movement. However, most expansion joints cause an interruption in the road surface which will, in turn, reduce the quality of ride and because of the continuous impact of traffic, tend to require fairly regular inspections and, in some cases, maintenance. The quality of ride and the maintenance required during the working life of the expansion joint are to a great extent dependent upon the initial quality of workmanship used during the placement of the joints. If expensive maintenance and in some cases complete replacement is to be avoided, good quality control must be exercised during construction. For such reasons it is good practice particularly on long structures such as elevated highways to use as few joints as possible allowing for large

movements in those joints rather than many joints accommodating smaller movements. For small movements, up to 25 mm, the expansion joint may be formed using a filler such as formed plastic or neoprene strips. Larger movements may also be taken up using neoprene strips or mechanical joints. The commonest form of mechanical joint is based on interlocking steel fingers which lie flush with the surface of the road. However, because of the cantilever action on the steel fingers caused by traffic the joint does require heavy fixings to the plates in position.

The ingress of uncontrolled water through expansion joints can lead to serious deterioration of the underside of the deck and its supports causing staining and more seriously spalling of concrete and corrosion of steel in the structure or in the bearings, see Figure 6.14. It is not only the water that causes the damage but the addition of salt used for de-icing, the chloride ions in the salt producing a particularly aggressive solution. It is, therefore, important to control the ingress of water by either using waterproof joints or if this is not practical to provide a system of drainage to take the water

Figure 6.14 Serious spoiling and staining to concrete cross beam

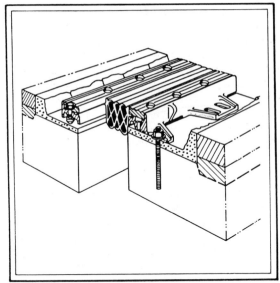

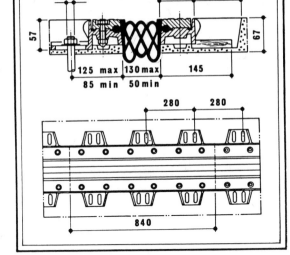

Figure 6.15   Freyssi expansion joint

away. When designing drainage details of this nature consideration must be given to access so that the drainage runs can be properly maintained. Figures 6.15 and 6.16 show some typical expansion joint details and placement.

Bridge construction

The construction of the sub-structure of a bridge, that is the foundations, abutments and piers, is typical of civil engineering work which if on land is fairly straightforward or if in water will require the use of cofferdams or caissons but can essentially be considered as general construction work. It is, however, the construction of the deck and in the case of arch bridges the arch, that may require constructional techniques that are often unique to this form of structure particularly for medium and long spans.

Short span reinforced and pre-stressed concrete decks are generally built using the traditional method of formwork supported on falsework. The main limitation of this method is that the falsework will close whatever gap the bridge is crossing. This leads to the point that if there are restrictions such as navigational or traffic flows to be maintained during construction, the deck type may be dictated to some extent by the form of construction required to keep the area to be bridged clear of obstructing falsework. For example, short span pre-cast concrete or steel sections can be positioned by a crane working off a pontoon or from the bank, the deck topping in the case of concrete

sections being cast on permanent formwork. Arch bridges if not built using the traditional method of centres supported off false work can be constructed using insitu or pre-cast segments by cantilevering out from the banks using temporary stays to support the segments or formwork. Figure 6.17 shows some typical stages of construction for an arch bridge using this technique.

Where medium to long pre-stressed or steel decks are constructed the method of segmental construction

Figure 6.16   Installing expansion joint

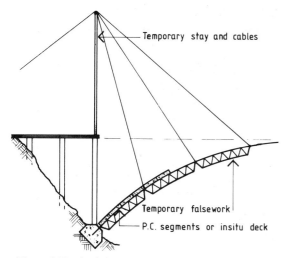

Figure 6.17 Arch construction

is often adopted. Segmental construction has been defined as a method of construction in which primary load supporting members are composed of individual members called segments post-tensioned or welded together. The segments are usually 3 m to 6 m in length but full deck width and depth in section, although in the latter case this is not always so. The principal methods of this form of construction are:

1 balanced cantilever;
2 span-by-span;
3 progressive placement;
4 incremental launching.

Although the following descriptions will tend to concentrate on concrete deck structures, the reader should realise that the techniques are with certain variations equally applicable to steel box segment decks.

1 Balanced cantilever

The balanced cantilever method of construction was developed to dispense with the need for falsework supported from the ground and as a method is applicable to both insitu and pre-cast concrete segments and steel box sections. To form a new segment for insitu construction a traveller, which is a moving formwork on rails, moves onto the last segment placed and is anchored at its rear end and aligned. Working from suspended cradles and in the forms the reinforcing steel and pre-stressing cable ducts are positioned and the concrete poured. After a suitable period, the prestressing cables are inserted and post-tensioned. The

complete cycle is approximately three to four days. In the case of pre-cast segments, the erection can be carried out from the deck using a beam and winch to lift the segments on a crane working from the ground or on a pontoon. The two methods require a special gantry fixed to the pier to cast, or place in the case of pre-cast segments, the first segment over the pier which is a difficulty that is overcome by a third method which entails using a launching gantry. Figure 6.18 shows the general principle of this method.

2 Span-by-span

The span-by-span method is an adaption of the balanced cantilever method developed for long structures with relatively short spans such as an elevated highway. The deck in this case is again constructed using either a traveller for insitu work on a beam and winch system, however, in this method the traveller or beam and winch is supported from a steel superstructure over a complete span. An alternative is to support the traveller from below on the ground or off the piers, the deck is constructed in one direction only. See Figure 6.19.

3 Progressive placement

The method of progressive placement was developed by French engineers and the order of construction is similar to that of span-by-span as the deck is constructed from one pier to the next. However, in this case the deck is cantilevered out. Used on spans of up to 90 m, the first third of the span is built out using post-tensioned segments as a free cantilever. To limit the large cantilever stresses, a temporary pylon is erected on the pier preceding the cantilevered section of deck and the final segments supported by continuous stays anchored back, via the pylon, to the completed span. On completion of the span the pylon is dismantled and the process repeated for the next span. The segments are delivered along the completed deck on rails and

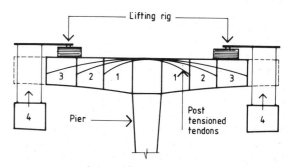

Figure 6.18 General principles of balanced cantilever construction

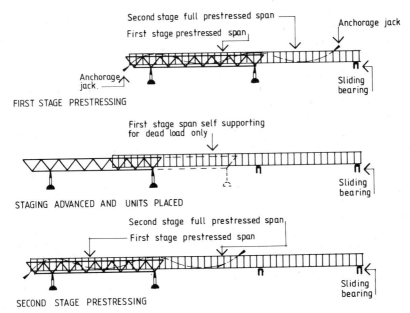

FIRST STAGE PRESTRESSING

STAGING ADVANCED AND UNITS PLACED

SECOND STAGE PRESTRESSING

Figure 6.19   Operating sequence for span-by-span construction of segmental precast structure

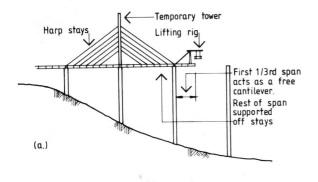

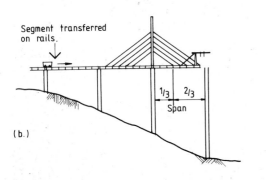

Figure 6.20   Progressive placement construction

positioned using a suitable crane. Figure 6.20 shows the basic principle of the method.

4   *Incremental launching*

Incremental launching or 'push out' construction as it is sometimes referred to was originally conceived by two German engineers, Baur and Leonhardt, in the early 1960s for the construction of the Rio Caroni bridge in Venezuela. In this method a casting bed for the segments is set up behind one of the bridge abutments and aligned so that the segments, in lengths of up to 30 m, are at deck level. The segment is then cast and allowed to gain a specified strength, it is then jacked forward on low friction sliding bearings and the formwork re-set so that the next segment is cast using the first segment as a stop end. The process of casting a segment and then jacking forward is repeated for the complete length of deck, each segment being post-tensioned to the adjacent segments. As the deck is pushed out it obviously acts as a cantilever until it reaches a pier, to reduce the length of cantilever one of two methods can be used.

The first method is to fix to the leading segment a steel launching shoe which projects out to reduce the cantilever distance between the spans, the second method employs a temporary cable-stay construction to support the first two spans. For long spans, the maximum so far being 100 m, where conditions allow such as for viaducts, temporary piers between the permanent piers may be erected. The bridge alignment for this form of construction must either be straight or a curve of constant radius and the deck of uniform depth. Figure 6.21 shows typical arrangements of this method.

The method of segmental construction is also adopted for cable stayed bridges as the stays give support to the bridge deck as it cantilevers out during construction. Figure 6.22 shows the part-constructed Adhamiyah bridge which crosses the River Tigris in Iraq. To minimise the pier width a single-plane cable system on the bridge centre line has been adopted. The cable system is of the harp form, the outer cable being anchored to the pier at the end of the bridge and the inner cable to a tie-down pier. The principal support member of the deck is a spine box girder 7.5 m wide and the tower is formed from shop fabricated steel box sections.

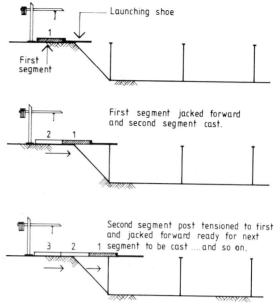

Figure 6.21   Incremental launching

Figure 6.22   Adhamiyah bridge, Baghdad, Iraq, under construction

7 Falseworks

Formwork

Formwork for concrete work could be described as a mould or box into which the freshly mixed concrete may be poured and compacted so that it will flow and finally set to take up the shape of the inner profile of the formwork or mould, it is important to realise that the inner profile must be opposite to that required for the finished shape and appearance of the concrete surface. To achieve this the general design and construction requirements are:

(a) The formwork should be durable and rigid and should be constructed so as to prevent bulging and twisting.
(b) The sheeting or facing against which the concrete is cast must be thick enough to withstand the pressures of wet concrete.
(c) The formwork must have sufficient strength to support the load, weight and pressure of wet concrete and also that of men, equipment and materials used in the concreting operation.
(d) Forms should be set correctly to line and level within the specified limits and adequately braced.
(e) The joints between formwork and previously cast concrete, and between the various members making up the formwork to the new pour should be sufficiently tight to prevent loss of mortar from the concrete, compaction of the concrete by vibration will expose any weakness in the formwork such as joints which are not well constructed.
(f) Forms and formwork must be of a reasonable size and weight for easy handling with the equipment available, the forms should be designed to be erected easily and speedily but they must also be capable of being struck in the required sequence – for example, side forms should be capable of being struck from beams leaving the soffit forms in place.

The types of formwork available for use on a project are now considerable. Apart from traditional timber which is still common, formwork materials include steel, plywood, polystyrene, hardboard, glass reinforced plastic and concrete, added to this form liners such as thermoplastic and other sheeting type materials may be used – for example, foamed polyurethane.

A number of proprietary formwork systems are available and on large contracts the contractor may opt for a system which can meet his needs. As a method of permanent formwork, materials such as pre-cast concrete, brickwork and woodwool slabs can be used.

Slipforming is a method of construction, used on structures such as silos, chimneys, liftshafts etc., and is economical for use on structures over 12 metres high, the slip form is made to slide continuously as the concrete placing proceeds without stopping, avoiding loss of time during striking and re-positioning the forms.

With slip forming the forms are lifted by hydraulic jacks operating on steel rods which are left in the concrete.

Falsework has been defined as 'any temporary structure used to support a permanent structure during its erection and until it becomes self supporting' – this definition covers the use of tubular scaffolding to support formwork for the construction of a bridge deck to the simple adjustable steel prop which is extensively used to support formwork from lintels to whole floor on roof pours.

Formwork materials

Timber and plywood are probably the most commonly used materials for formwork since they are easy to cut and erect on the site.

When using either plywood or timber planking the material is made up and backed by a frame to a size which can be readily handled by the available equipment on site. For very large pours with multiple usage of the forms timber framing with a plywood face is more economical, great care is needed during the assembly process so that no damage is caused to the plywood face since this may be reproduced as a blemish of the finished concrete surface, and when striking formwork similar care is needed so that no damage is caused to the newly cast concrete.

Steel formwork is used in proprietary systems as panels with either a steel or plywood face on metal framing supported by tubular scaffolding or a similar system, these systems can be used for walls, floors, columns or beams and, providing they are properly cared for, many more uses can be obtained from them than from traditional timber or plywood formwork, the systems may be hired from plant hire companies or bought in by a contractor, see Figures 7.1 and 7.2.

Figure 7.1 Patent steel deck forms

Other materials

Holes which were previously formed from plywood and timber can now be formed from expanded polystyrene as can formers for hollows and voids in bridge decks, but these materials can only be used once and must be properly secured to avoid displacement and distortion.

Waxed cardboard is used for forming circular columns, and oil tempered hardboard may be used as a facing for curved sections of formwork, glass reinforced plastics and vacuum formed plastic facings can be useful when complicated surface features and odd shapes are required, but care has to be taken when compacting the concrete that no damage is caused when vibrating in order to compact the wet concrete.

Release agents are painted onto formwork to enable the forms to be struck more easily from the newly concreted face, different types of formwork require a different release agent and therefore it is important to ensure that the correct one is used. Care must also be exercised to ensure that the release agent does not contaminate the reinforcement. Details of various release agents are shown in Table 7.1.

Figure 7.2 Patent steel wall forms

Table 7.1 Some release agent types

Release agent type	Comments
Neat oil with surfactant	A useful general purpose release agent for all types of formwork, including steel. Over-application may result in staining of the concrete. Oil film may be affected by heavy rain.
Mould cream emulsion (oil-phased)	Widely used release agent recommended for all types of formwork except steel. Especially recommended for absorbent forms such as timber. Suitable for high-quality finishes. Mix thoroughly before application and use as supplied without further dilution. Avoid using emulsions when there is a risk of freezing. Storage life may be limited.
Chemical release agent	Recommended for all types of formwork. Suitable for high-quality finishes. Based on light, volatile oils which usually dry on the surface of the form to leave a thin coating which is resistant to washing off by rain. The dried coating gives a safer surface to walk on than an oily film and the release agent does not then transfer from operatives, footwear onto reinforcement. Rate of coverage is greater than for conventional oils, also more expensive for a given volume but can be economical if used sparingly.
Wax	Recommended for moulds made of concrete. Difficult to apply a thin, uniform coating.
Barrier paint	Not recommended for use without a release agent because the paint film on its own would soon become scratched and lose its release capacity.
Neat oil (without surfactant)	Not to be used where appearance is important; encourages the formation of blowholes. Inexpensive and can be used where concrete is later to be covered.
Water-phased emulsion (oil-in-water)	Not to be used for visual concrete. Causes severe retardation and discolouration. Cheap, easy to apply and can be used if concrete is not exposed to view.

Beam and slab forms

Formwork to beams sides is usually supported by the soffit form so that the sides may be struck before the soffit which has to be left in place until the beam can support its own weight, slab formwork is usually supported by that of the beam where a beam is incorporated into the design.

Wallforms

Wallforms are held together with wall ties which pass through the timber framing on each side and, in the case of large wall formwork panels, the formwork is built-up plywood sheeting fixed to timber framing and supported by pairs of steel channels which act as walings, the wall ties will pass at intervals between the pairs of channels and if another lift to the wall is to be concreted the position of the wall ties at the top of the first lift will be common to the position of the ties to the second lift, see Figure 7.3.

With most wall formwork it is necessary to provide bracings by means of timber struts or steel props to keep the formwork aligned and to resist pressure from the concrete being placed.

Column formwork

Typical formwork for columns is made from plywood sheeting with timber bearers spaced at intervals along the length of sheeting, two sides of the column will have formwork of a width equal to that of the column while the two opposite sides will be the column width plus an overlap equal to the plywood sheeting thickness. The

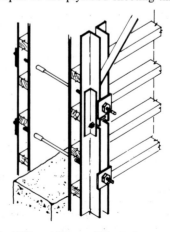

Figure 7.3   SGB metriform wall shuttering

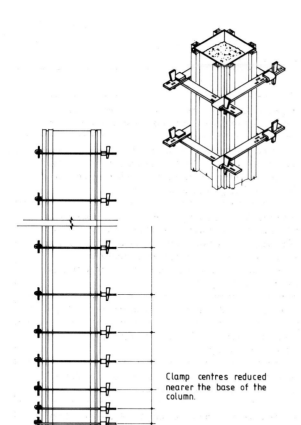

Figure 7.4 Column formwork

panels are held together by adjustable steel clamps which are fixed around the column on the backing bearers after erection, the spacing of the clamps should be such that near the bottom of the column formwork a closer spacing is required to resist the hydrostatic pressures which build up on the formwork by the placing and compaction of concrete as the column is concreted to its full height, see Figure 7.4.

Circular columns may be formed by using proprietary

Table 7.2 Minimum periods before striking formwork

Formwork	Surface temperature of concrete	
	16°C	7°C
Vertical forms to columns, walls and large beams	12 hours	18 hours
Slab soffits	4 days	6 days
Beam soffits (props left under)	10 days	15 days
Props to slabs Props to beams	10 days 14 days	15 days 21 days

steel forms, plywood or hardboard sheeting, glass reinforced plastic column moulds or waxed cardboard moulds.

Striking and storage

Formwork must be designed so that it may be struck without damage to the concrete and in the case of large panels with the safety of operatives engaged on the work in mind, Table 7.2 gives guidance to the striking times when ordinary Portland cement is used. As soon as the formwork has been struck it should be cleaned to remove grout and dust and in the case of steel forms it should be lightly oiled before storage, panels and plywood sheets should be stored horizontally on a flat level base. Large panels are best stored vertically with the concrete faces together, small items are best stored in boxes, and soldiers and walling which are loose should be stored with their forms to prevent loss.

Scaffolding

Traditional scaffolding is still the most widely used method – however, several modern systems are now available which are more easily erected than traditional scaffolding. Some have couplers already fixed to the tubes, others use special locking devices so that couplers are not needed at all, these modern systems however are all of the independent scaffold type.

Scaffolding has the advantage of being extremely versatile, enabling it to be erected where it would be impossible to stand an independent or mobile scaffold tower such as stepped or sloping ground, or it may be cantilevered out from the permanent structure or when designed to do so may be used as a support to falsework or formwork for large concrete pours, see Figure 7.5.

Putlog scaffold (Figure 7.6)

With this type of scaffolding there is only one row of standards placed about 1.3 m from the wall. The cross members called putlogs are supported on the wall as building proceeds or are inserted into mortar joints which are raked out to receive them. Putlogs have a flattened end which can be inserted into the mortar joint, and this type of scaffolding is very easily extended as work proceeds.

Independent scaffold (Figure 7.7)

This type of scaffolding stands clear of the building as the name implies. It is built using a double row of standards set about 1 m apart with the inner row between 100 mm to 530 mm from the wall, scaffold boards are then placed on cross members called transoms to form the working platform.

Figure 7.5   Scaffolding used to support formwork

Figure 7.7   Independent scaffold

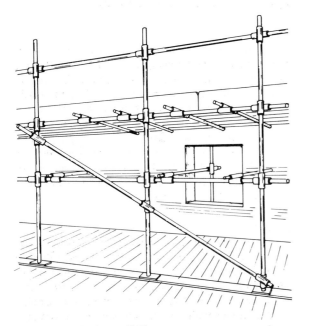

Figure 7.6   Putlog scaffold

Mobile and static towers (Figure 7.8)

A mobile tower is a scaffold mounted on wheels with a single working platform, used mainly for light work such as painting or for simple maintenance, and its height is governed by the smallest base dimension (SBD). For internal towers the maximum height is 3.5 × SBD and for an external tower the maximum height is 3 × SBD with a total maximum height of 12 m unless specially designed, the smallest plan size permitted for a mobile tower is 1.2 × 1.2 m but the tower shape may be square or rectangular. Handrails and toeboards must be fitted and a minimum gap must be left for access in the handrail if a ladder is fitted to the outside. Access to the tower may also be gained by fixing the ladder to the inside which may improve the stability of the tower.

Static towers may be used as working or observation platforms for maintenance services or sporting events. Foundations capable of carrying the desired load must be provided, baseplates must be provided under all standards, if the tower exceeds 10 m in height it must be adequately tied to the main structure, bracings must

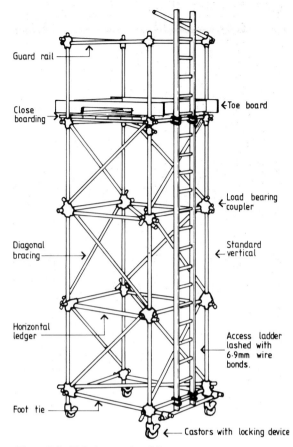

Guard rail

Close boarding

← Toe board

Diagonal bracing →

Load bearing coupler

Standard vertical

Horizontal ledger

Access ladder lashed with 6·9mm wire bonds.

Foot tie

← Castors with locking device

Figure 7.8 Tubular metal mobile tower

of course be used to keep the tower rigid. If the tower is independent of any permanent structure or building it should be tied by one of the following methods:

(a) By the use of guy wires fixed to the tower every 10 m vertical and at an angle of 45° to the anchorage points.
(b) By holding down the bottom corners of the tower.
(c) By the use of adequate weights at the base of the tower.
(d) By constructing buttresses at the base of the tower.

The working platforms should be close boarded and for towers exceeding 2 m, ladders, handrails and toeboards must be provided. Other static towers that may be found on construction sites are hoist towers which are erected to support the guides and masts of mechanical hoists, the tower must be designed to suit the particular make of hoist and the loads it will have to carry.

Parts of a scaffold

Tubes

Usually made of galvanised steel or an aluminium alloy which is lighter and more easily handled but the aluminium can suffer from corrosion if it comes into contact with cement, lime or salt water. The tube lengths range from 1.8 to 6.4 m in length.

Standards

These are vertical tubes in a section of scaffolding. Their spacing depends upon the use the scaffold is to be put to and the load to be placed upon the scaffolding. Normally they will be placed 2.5 m apart to form a standard size bay which will support two men, their mortar and about 80 bricks. When scaffolding is being built fairly high the tubes will have to be joined to make the standards long enough. It is important in this case to ensure the joints between adjacent standards are staggered to produce a stronger structure.

On putlog scaffolding the standards are usually placed about 1.3 m from the wall which gives space to place five boards wide as a working platform and space between the edge of the inner board and the wall of 100 mm. With an independent scaffold, the clearance from the inner row of standards to the wall will depend on what the scaffolding has been erected for. If access to a roof is the purpose the standards should be within approximately 100 mm, for bricklaying the standards would need to be 350 mm away from the wall allowing one board to be placed on the transom which will project past the inner row of standards.

Ledgers

These are horizontal tubes which link the standards together along the building and on putlog scaffolding they should be fixed every 1.35 m high so that putlogs can be rested on the ledger and built into the brickwork, this spacing is important to achieve level putlogs on which the boards are to rest. On independent scaffolds, the ledgers can be fixed up to 2.6 m apart and a 2 m spacing will have enough room to walk through the scaffold. Modern systems with locking devices or couplers already fixed to them use standard size ledgers to fit either 1.8 or 2.5 m bays.

Putlogs

These are the short tubes which form the cross members on a putlog scaffold and they provide support to the scaffold boards. The putlog tube has one end flattened so that it can be built into the horizontal brickwork joint of the wall under construction, and there is a scaffold fitting available that will convert an ordinary scaffold tube into a putlog.

The spacing of the putlogs should be at a maximum of 1.5 m but this spacing may have to be reduced to ensure that the boards are supported at their end avoiding traps in the boards. A board should not overlap a support by more than four times its thickness giving an overlap of 150 mm for a 38 mm thick scaffold board.

Transoms

These are the short lengths of tube forming the cross members on an independent scaffold. They carry out the same function as a putlog and are spaced at the same intervals but they are attached at both ends to the ledgers and are not supported by the wall. In modern scaffolding systems additional intermediate transoms can be hooked over the ledgers if additional support is needed.

Braces

These are lengths of tubing fixed diagonally to stabilize the scaffold structure. All types of scaffolding require longitudinal bracing fixed across the face to prevent collapse, with traditional scaffolding a brace should be fixed to the bottom of every fifth standard and modern scaffolds with their locking systems of fixing should have one in every eight bays braced. This bracing must be taken to the full height of the scaffolding and a minimum of two bays must be braced if the scaffold is more than four bays long. Traditional scaffolding also requires transverse braces fixed at regular intervals to avoid the scaffolding falling away from the building.

Guard rails

These can be lengths of tubing fitted to the standards, and they are essential for safety and must never be omitted. They should be fixed at a height of between 915 mm and 1145 mm above the working platform and the space between the top edge of the toeboard and the guard rail should not exceed 760 mm, this space is filled by hanging a mesh type safety fence on the guard rail to be fixed at toeboard level.

Toeboards

These are scaffold boards on edge and should always be fitted as they prevent tools, bricks and any other material from being accidently kicked off the platform; they also prevent anyone slipping off the edge.

Base plates

These are usually square metal plates with a central spiggot that fits inside the base of the standard to prevent the standards sinking into the ground. Adjustable

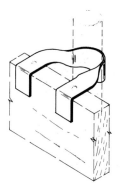

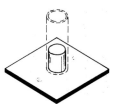

Figure 7.9 Baseplate

baseplates are needed on sloping ground and on soft ground boards called sole plates should be placed under the baseplates to spread the load more evenly, see Figure 7.9.

Couplers

These are needed on traditional scaffolding to join the various tubes together. There are several types (see Figures 7.10, 7.11 and 7.12) each doing its own job:

Double couplers
Putlog couplers
Swivel couplers
Sleeve couplers
Final couplers

Fans and nets

It is often necessary on sites to provide protection of the general public or adjoining property from materials falling from the scaffold. This can be achieved by the erection of fans or nets on the outside of the scaffold.

The fan consists of support tubes extended from the scaffold fixed across two ledgers and extending outwards from the scaffolding. Various materials are used for covering the fans, the most common are scaffold boards or corrugated iron sheets. Fans should normally be erected as close to the working platform as possible.

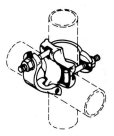

DOUBLE COUPLER

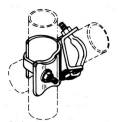

SWIVEL COUPLER

PUTLOG COUPLER

Figure 7.10 Double and putlog couplers

SLEEVE COUPLER

Figure 7.11 Swivel and sleeve couplers

On some construction sites a fan is erected at the base of the scaffold and others erected as the work proceeds. It may in some cases be necessary to support the fans from below and this can be achieved by fixing raking scaffold tubes from those supporting the fan covering back to the scaffold itself. It should be emphasised that fans are for protection only, they should not be used as a place to store materials and warning notices to this effect should be displayed on the fans.

Safety nets are provided to prevent death or injury to persons who fall or to protect persons or property from objects which fall. The nets should be suspended at or below places where operatives or materials may fall and where other precautions are impractical, or when nets are considered to be an essential extra.

Figure 7.12 Final coupler

8 External works

Construction of sewers

When the line, levels and gradient of the sewer have been established, the position of the manholes has been decided upon and the materials specification has been finalised, then the work of constructing the sewer may begin.

Before the pipes can be laid to a true line and level it will be necessary for the trench to be excavated. To do this accurately with a bottom accurate to the falls shown on the drawings it is essential to set up sight rails to the line and gradients shown on the drawing and ideally painted black and white. A traveller should then be prepared and cut to the depth of the pipe invert plus the depth of bedding of the pipe and to this length must be added the height that the sight rails have been set above ground level, see Figure 8.1.

During excavation care should be taken to ensure that the excavated material is placed away from the edge of the trench to reduce the dead load on the vertical sides of the excavation and avoid possible collapse. If there is any doubt at all that the trench sides may not be self supporting then they should be timbered to avoid danger to operatives working in the trench or along the line of excavation. In any event, a trench deeper than 1.2 m in soil should be timbered.

Having completed excavation between manholes the pipe laying can begin, the pipes may be laid on a concrete bed or on a granular material depending on the specification.

Pipes with rigid joints are laid on a concrete bed which is either scooped out to receive the collars or tamped true to the falls, so that the pipes are supported on the collars. The pipes are then either carefully haunched or surrounded with concrete.

Pipes with flexible joints are usually laid on a granular bed and surrounded with a granular material, see Figure 8.2, which should be well compacted on either side of the pipe. Pipes are laid to a true line by drawing a taut string line either along the top of the pipeline, or to one side of the pipes.

Whichever method is employed the pipes must be laid to a true line and gradient beginning at the lower manhole level and working towards the head of the sewer.

With rigid pipes, the spigot end of the pipe is first wrapped with gasket or hemp and then placed into the socket of the pipe previously laid, then by the use of a caulking tool and hammer, the gasket is caulked into the joint of the pipe. It is then checked for line and level before sealing with a mortar of 1:2 cement/sand and finishing in a fillet around the outside of the pipe. In some cases, it may be found more convenient to lay a number of pipes before sealing the joints with mortar, see Figure 8.2.

On flexibly jointed pipes which is now the most common method, laying techniques are speeded by the use of 'push fit' type joints which do not require caulking or sealing with cement mortar. Because the pipes are supplied in longer lengths the number of joints can be reduced. These flexible joints are available with a number of different materials such as vitrefied clay, concrete, asbestos cement, PVC and pitchfibre. The type of joint and methods of jointing vary from one material to another and information is provided by the manufacturers on the correct method of jointing them.

On completion of the laying operation and before surrounding the pipeline and back filling the trench the sewer line should be tested, to ensure that it is capable of carrying out the work for which it has been designed. A number of testing methods exist. The choice of test depending on the nature of the liquid to be carried by the pipeline.

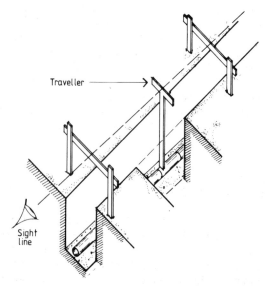

Figure 8.1 Typical arrangement in the use of sight lines and traveller

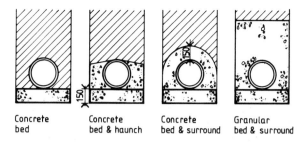

Concrete bed | Concrete bed & haunch | Concrete bed & surround | Granular bed & surround

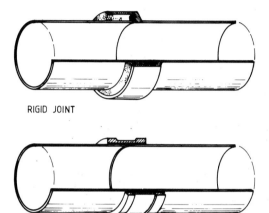

RIGID JOINT

FLEXIBLE JOINT

Figure 8.2 Pipe bedding and jointing

(i) Water test

It is the most logical test since the sewer will eventually carry liquid. In this test a stopper or bag is inserted in the outfall end of the pipeline and the line filled with water to provide a head of water of between 1.2 and 6 m. An appropriate period should be allowed for absorption of water by the pipes, then the loss of water over a period of 30 minutes should be measured by adding water from a measuring vessel at regular intervals and noting the quantity required to maintain the original head. CP2005 (1968) sets out the criteria for the test.

(ii) Air test

This type is considered a more severe test than the previous method, but it is a more convenient way of testing drains and sewers before back filling. CP2005 (1968) requires that the length of pipeline under test should be effectively plugged and air pumped in by a suitable means, e.g. a bellows or air pump, until a pressure of 100 mm in a glass U tube connected to the pipeline is indicated. The air pressure should not fall to less than 75 mm during a period of five minutes.

(iii) Smoke tests

A smoke cartridge is introduced into the line of pipes to be tested and then both ends of the line are plugged. An air line is then used to increase the pressure in the line of pipes under test. As an alternative, a smoke making machine may be used in which cotton waste or similar material is burned in a chamber to produce smoke which is then pumped into the line of pipes by a bellows which forms an integral part of the machine.

(iv) Light test

Other tests may include a light test which employs the use of a light at one end and a mirror tilted at 45° to the line of pipe at the other manhole. If the pipeline is free from obstruction and laid true a clear image of the light should appear in the mirror.

A **steel ball** is sometimes used to run down a line of pipe to check that the line is free from obstruction, the ball is 13 mm smaller than the diameter of the pipe.

On existing sewers where it is thought there may be an obstruction or other problem, a **television survey** may be carried out.

Road gullies

Rain falling on the carriageway and footways of the road network is collected into the channels and discharged through gullies into the sewer system. Gullies are usually spaced so that their maximum catchment area is approximately 190 m^2, the exact spacing depends on the width of road and footways and whether the road is cambered or is a straight crossfall. In a section of road with a straight crossfall, gullies are only required on the lower channel. Gulley gratings and frames are manufactured to BS497 (1976) and may be sited either in the channel or as side entry gullies with the gulley face conforming to the profile of the kerb, see Figure 8.3.

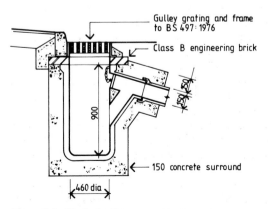

Figure 8.3 Typical gulley arrangement

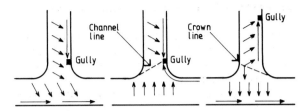

Figure 8.4   Drainage at junctions

Gully pots should conform to either BS556 Part 2 (1972) *Concrete* or BS539 (1971) *Salt Glazed*, they may also be manufactured from UPVC for which there is no BS at present, or be built insitu of brickwork. No attempt should be made to cut costs by omitting gullies and where gradients are steeper than 5% or flatter than 0.5% additional gullies should be provided. The gullies should be sited just uphill of the tangent point at road junctions so that surface water in the channel does not flow across the junction, see Figure 8.4.

Manholes

Manholes can be divided into two classes, simple manholes for use on small sewers, and special manholes on large and usually deep sewers. These include backdrop manholes, side entrance manholes, storm overflows and pressure manholes. This class of manhole construction is expensive to build and was in the past almost invariably built in brickwork.

Mass concrete, precast concrete rings and reinforced concrete are now the materials which are in common use and cheaper to construct than traditional brickwork. A combination of these materials is often used. Figures 8.5, 8.6, 8.7 and 8.8 illustrate some of the types of manholes in use.

(i) Brickwork manholes

These are usually constructed as a rectangular chamber on a concrete foundation with either a reinforced concrete roof slab or a brick arch, and an access shaft 675 x 750 mm in size. Class B engineering bricks to BS3921 (1974) laid in English bond on a 1:3 cement/sand mortar bed and pointed flush are used to construct the manhole. In good ground 225 mm brickwork is adequate for depths up to 3 m but an increase in wall thickness will be necessary for extra depth, or unusual ground conditions, or to resist water pressure. Step irons are built into the brickwork on a 300 mm staggered line. But for depths greater than 5 m galvanised iron access ladders are provided.

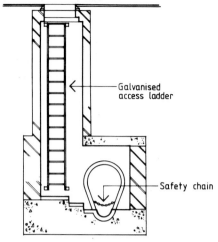

Figure 8.5   Side entrance manhole on large sewer

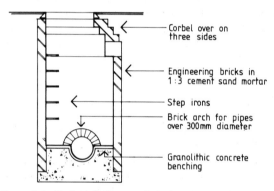

Figure 8.6   Typical shallow manhole

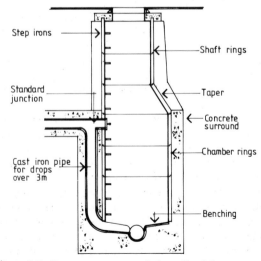

Figure 8.7   Precast concrete manhole and backdrop

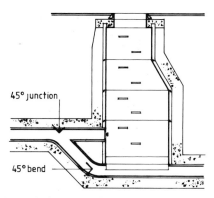

Figure 8.8 Typical ramp manhole up to 1.5 m drop

(ii) Precast concrete ring manholes to BS556 (1972)

These may be built on a precast or insitu concrete base. They are constructed rapidly using unskilled manpower. This type are particularly effective in poor ground conditions and are surrounded with 150 mm of concrete to ensure watertight conditions. Taper sections are available to reduce the chamber size to a 675 mm diameter shaft. Step irons are provided in the precast concrete ring sections, but manholes deeper than 5 m are provided with access ladders of galvanised iron.

Manholes are provided at all changes or direction, gradients, and pipe size, at junctions and at heads of all sewers. The maximum distance apart of manholes for small sewers should be 30 m, although this distance may be increased for sewers over 1 m diameter.

Connections to local authority sewers

The Public Health Act 1936 and the Building Regulations lay down the legislation for connecting drains to Local Authority sewers. The methods of connection depend upon the position of the drain in relation to the sewer, the size and condition of the existing sewer, the material the sewer is constructed from and the difference in invert levels.

Whichever method is chosen the drain entering the sewer must discharge its contents obliquely in the direction of flow of the sewer, and the connection must be made so that it will remain watertight under all flow conditions and will work satisfactorily.

The methods of connection may be an existing or new manhole where the gradient of the contributing drain can be increased to allow the connection to be made at invert level, or by the use of a ramp if the invert level difference is less than 1.50 m. If it is not economical to increase the gradient of the new drain

and a ramp will not fulfil the function, a new backdrop manhole may be constructed.

In the event of a new sewer being constructed to serve an area 'stopped' junctions are provided in the line of the sewer to allow connections to be made from existing properties.

If a new property were to be built that had not been allowed for, then on a small sewer up to 225 mm diameter serving an area, an oblique junction could be inserted in the line by removing two or three pipes and replacing one of the straight pipes with the junction, then connecting the new property to this junction, see Figure 8.9.

To connect a property to a larger sewer then a saddle connection, see Figure 8.10, is made by carefully breaking a hole in the top half of the sewer pipe and trimming the opening so that the saddle pipe fits neatly onto the sewer pipe, it is then jointed up with cement mortar and surrounded with concrete.

The dangers to operatives working in sewers is greater than is generally realised. The risks to which the sewer worker is exposed can be summarised as follows:

(a) Falling from ladders or step irons;
(b) Injury from tools being dropped down a manhole;
(c) Injury by falling while travelling in a sewer;
(d) Being swept away or drowned;
(e) Infection (Weils disease), Leptospiral jaundice;
(f) Poisoning by gases or vapour;
(g) Physical injury by explosion.

The way to minimise accidents in sewers is to train operatives in safe working practice, to recognise potential hazards, and in the case of an accident the

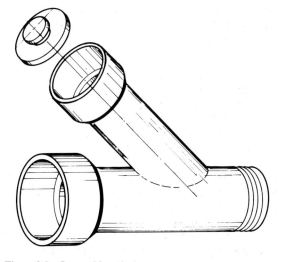

Figure 8.9 Stopped junction

Figure 8.10   Saddle connection

procedure to be followed. The safety precautions to be exercised in sewers are beyond the scope of this text, but may be found in the Local Government Training Board's book *Safety in Sewers* and further reference may be made to the Institution of Civil Engineers' booklet *Safety in Sewers and Sewage Works*.

Surface water run-off and discharge

Providing adequate surface water drainage facilities are an essential part of any road construction scheme. It must cater for the run-off from the carriageway, footway, verges and hard-shoulders on motorways, and for the run-off from catchment areas adjacent to the road line, which may be affected by the construction.

Road Note 35 (1976) *A Guide for Engineers to the Design of Storm Sewer Systems* sets out recommended guidelines for the design of drainage systems for small areas such as housing estates where the largest sewer will probably not exceed 600 mm diameter, and the Transport and Road Research Laboratory Hydrograph method which will provide accurate sewer design for urban areas.

Before surface water calculations may be made, a number of terms on which these calculations are based should be understood:

(a) Catchment area
The total area from which run-off of all surface water would flow by gravity to a sewer.

(b) Time of concentration
The time taken for water to reach the point under consideration after falling on the most remote part of the surface, its value is given by the sum of the time taken to flow across the surface and enter the sewer (time of entry) and the time to flow along the sewer assuming full bore velocity.

(c) Time of entry
It is recommended that a time of entry of two minutes should be used for normal urban areas increasing up to four minutes for areas with large paved surfaces and slack gradients.

(d) Time of flow
This is calculated assuming full bore velocity.

(e) Intensity of rainfall
This depends on storm duration and frequency of storm and it is appropriate to use a mean rate of rainfall during a storm. The duration of the storm should be taken as being equal to the time of concentration of the drainage area, to the point for which the calculation is being made. In the absence of precise local data, the Ministry of Health formula may be used for most road drainage systems:

$$R = \frac{750}{t + 10} \text{ mm/hr for storms of 5–20 mins duration}$$

$$R = \frac{1000}{t + 20} \text{ mm/hr for storms over 20 mins duration}$$

Where t = time of concentration

It is suggested in Road Note 35 that the following is adopted when using the Rational (Lloyd Davies) formula for the calculation of surface water sewers up to 600 mm diameter.

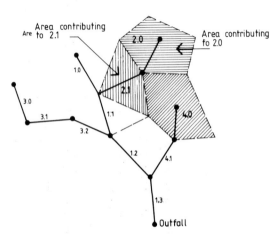

Figure 8.11   Example of decimal classification

Example of pipe length 1.1

CRIMP AND BRUGES VALUE (   )

RATIONAL FORMULA DESIGN SHEET
Storm frequency one in 1 year
Time of entry 2 min/roughness coefficient 0.6 mm

1	2	3	4	5	6	7	8	9	10	11	12	13	14	15
								Impermeable area (ha)						
Pipe length (number)	Diff. in level (m)	Length (m)	Gradient (1 in)	Velocity (m/s)	Time of flow (min)	Time of concentration	Rate of rainfall (mm/h)	Roads	Bldgs. yards etc.	Total (9 + 10)	Cumulative	Rate of flow (l/s)	Pipe dia. (mm)	Remarks
1.0	1.10	63.1	57	(1.25) 1.33	0.79	2.79	67.9	0.089	0.053	0.142	0.142	(22.8) 26.8	150	
1.1	1.12	66.1	59	(1.61) 1.70	0.65	3.44	62.5	0.077	0.109	0.186	0.328	(66.2) 56.9	225	
1.2	0.73	84.7	116	(1.40) 1.46	0.97	4.41	57.4	0.081	0	0.081	0.409	( )		

Length number 1.1 Assume pipe size 150 mm dia
Length 66.1 – Difference in level 1.12 m
Gradient $L \div F$ 66.1 ÷ 1.12 = 1 in 59   Column 1
                                                Columns 2 and 3
                                                Column 4
Velocity of flow from tables 1.70 (1.61 Crimp & Bruges)   Column 5
Time of flow 66.1 ÷ 1.70 ÷ 60   = 0.65   Column 6
Add time of entry   = 2.00
Add time of flow of 1.0   = 0.79   Column 6
Total time of concentration so far   = 3.44 min   Column 7

$$Q = \frac{Ap \times 1}{0.360} = \text{litres/sec}$$

$$Q = \frac{.328 \times 62.5}{0.360} = 56.94 \text{ litres/sec}$$

Table 8.1   Design sheet with worked example (*From Road Note 35*)

(i) A key plan of the proposed sewer system should first be prepared, see Figure 8.11, using a decimal classification. In this way if it becomes necessary to add a further branch it does not matter as far as the calculation is concerned.

(ii) Prepare a design sheet to simplify calculations. Refer to Table 8.1.

(iii) Enter the basic design data on the design sheet, columns 1, 2, 3, 4, 9, 10 and 11, the total in column 11 is the area of surface directly connected into each length and is given by the sum of columns 9 and 10. Column 12 can now be completed by adding in the contributions from the branches at the appropriate junctions. This column gives the total area of surface contributing to the flow in a length. Reference to Road Note 35 will clarify the method of entry of the branches of the sewer onto the design sheet.

(iv) A pipe size (column 14) is now assumed, the pipe full velocity of flow (column 5) is found from published tables* and the time of flow (column 6) calculated from columns 3 and 5. The time of concentration (column 7) is the total time of flow up to and including the length under consideration plus the time of entry, say, two minutes where sewers join. The time of concentration is taken to be the greatest time to the manhole concerned.

(v) The rate of rainfall (column 8) corresponding to the time of concentration is found from tables published by the Meteorological Office or an approximation may be obtained from the Ministry of Health formula.

(vi) The expected peak rate of flow in the pipe is then given by the 'rational' formula:

$$Q = \frac{Ap \times i}{0.360} \text{ m}^3/\text{second, or } Q = \frac{Ap \times i}{0.360} \text{ litres/second}$$

where Q = run-off litres/second or m³/second
       Ap = impermeable area in hectares
       i   = intensity of rainfall mm/hour
using the figures given in columns 8 and 12.

The assumed pipe size is then checked to see if it can carry the expected flows, if not a larger pipe is assumed, and the steps from (d) are carried out again until a pipe of sufficient size is reached.

Tables for the hydraulic design of storm drains, sewers and pipelines, 2nd Edition.
  Crimp and Bruges, *Tables and diagrams for use in designing sewers and water mains*.

Construction of flexible and rigid pavement and methods of surfacing

Flexible pavements are a development of the Telford or Macadam principle of construction. They are of a layered construction which rely on the interlock of the particles in each layer for stability, and on the load spreading properties of the layers to distribute the wheel loadings to the subgrade.

Rigid pavements are of concrete slab construction, usually resting on a layer of granular material. They distribute the wheel loadings over a wider area and are often used on the weaker subgrades. The choice of construction is influenced by a number of factors which are taken into account at the design stage. These include the cumulative traffic flow and the expected growth rate over the design period, the assessment of subgrade strength and stiffness and the level of the water table below the formation. A few years ago about 90% of the new road construction in this country was of the flexible type, but since the oil crisis more construction of the rigid type of pavement has taken place.

Currently road design is based on Road Note 29 first published in 1959 and last revised in 1970. During this period there has been a dramatic increase in the number and size of heavy vehicles using our roads giving rise to concern about premature failure of motorways. In 1970 new motorways were expected to carry 11 to 20 million standard axles (MSA) during their life, now new and reconstructed motorways are required to carry 50 to 150 MSA and the M1 widening in Hertfordshire has been built to withstand 370 MSA. Now a new approach has emerged from the Transport and Road Research Laboratory in its report, LR 1132, on the structural design of bituminous roads, it defines design life as the length of time which 85% of roads are likely to survive before reaching their critical condition and need a strengthening overlay, the TRRL has then done economic evaluations of various possible design lives for different levels of road from lightly trafficked to busy motorways.

The Department of Transport is expected to publish a new standard to replace Road Note 29 doubling the design strengths for road pavements but whether the new standard will affect concrete pavements remains to be seen.

A glossary of highway engineering terms can be found in BS892 (1967). A number of the terms directly related to flexible and rigid construction in Figure 8.12 are described below.

Road pavement
The total depth of construction resting on the sub-grade which will support the traffic loads.

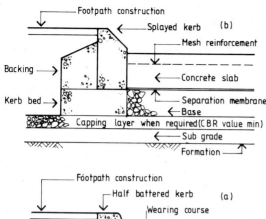

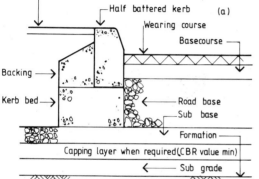

Figure 8.12 Flexible (a) and rigid (b) construction

Sub-base
A secondary layer of material provided between the formation (or capping layer) and the base or concrete slab in rigid construction.

Road-base
The layer which provides the principal support for the surfacing. In rigid construction this layer is the concrete slab.

Sub-grade
The natural foundation or fill which receives the loads from the pavement.

Formation
The prepared surface of the sub-grade on which the pavement is constructed.

Capping layer
When the CBR is less than 5% the Department of Transport normally requires the use of a suitable low cost capping layer. The capping layer is designed to

provide a working platform on which sub-base construction can proceed with minimum interruption from wet weather and to minimise the effect of a weak sub-grade on road performance.

Surfacing
The top layer or layers of the pavement comprising:
(a) Wearing course – the top layer of the surfacing which carries the traffic.
(b) Base course – the layer sandwiched between the wearing course and the road base. It is not always necessary to include a base course since surfacing may be laid as a single course.

Any reference to other terms which are not clear to the reader may be found in BS892 (1967).

The quality of the sub-grade is the principal factor in determining the thickness of the pavement. Deterioration by frost action must also be taken into account at the design stage since this could have serious consequences on the bearing capacity of the sub-grade. The strengths and stiffness of the sub-grade is assessed on the California Bearing Ratio (CBR value, a test developed in the USA). The test is described in full in BS1377 (1975) *Methods of testing soils for civil engineering purposes.*

The testing apparatus is of the type shown in Figure 8.13. The samples of soil must be compacted to a dry density similar to that which it is expected to achieve in practice on site. The sample having been prepared to simulate the maximum moisture content likely to be experienced at the completion of the road works in the

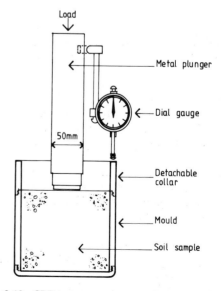

Figure 8.13 CBR test apparatus

sub-grade. The test results are plotted on a load penetration graph, Figure 8.14, and a line is drawn through the points plotted. The curve is usually smooth towards the zero point, but occasionally a correction has to be made (Test 2). If the line of curve from the zero mark is concave, this entails drawing a tangent to the curve at the steepest point of the curve, and projecting the line down to the penetration axis. At the point where this line cuts the axis a new point of origin is placed for the test. The loads required to produce penetrations at 2.5 mm and 5 mm are recorded and expressed as ratios of the loads required to cause the same penetrations in a standard crushed rock material.

Road Note 29 and TRRL Report 1132 recognise that roads will have to be built over relatively weak sub-grades. A table which correlates between CBR values and soil types based on British experience with a wide variety of sub-grades and moisture conditions for high and low water tables is shown in Table 8.2. The soils for pavement design are classified into seven main types, Table 8.2, and these fall into two categories:

(a) Cohesive soils (plastic) – which are clays and mixtures of clay and sand or silt.
(b) Non-cohesive (non-plastic) soils – which are sands and gravels, coarsely grained materials.

Coarsely grained soils are not generally frost susceptible, since the voids of the compacted material are sufficient to take up any expansion of soil and water due to freezing. Not all cohesive soils are susceptible to frost action, those which have fine grains (particles finer than 0.002 mm) are impermeable and therefore not susceptible. Frost susceptible materials usually have at least 10% to 15% of particles by weight smaller than 0.02 mm silts, fine sands and some chalks. To reduce the risk of frost damage to sub-grades, it may be necessary to increase the constructional depth of the pavement by at least 450 mm above formation.

Roads tend to be built to maximum gradients of approximately 1:25 (4%). This may involve a certain amount of balancing of the earth works between cut and fill as well as founding the road at existing ground levels. In this process soils are excavated at one point of the road line, where there is a surplus of material and transported to, spread and compacted at another point where the original profile of the ground is lower. Compaction of the soil reduces the air voids in the soil, and so reduces the risk of moisture change in the sub-grade. If the compaction is not carried out correctly during the earth works stage of the contract, then traffic using the road at a later date will cause further compaction to take place which will distort the road pavement. The information (Table 8.3) taken from the Department of Transport Specification for Road and Bridge Works lists the type of compaction plant, its category and the minimum number of passes for the compacted depth of soil type. Comparative field density tests carried out in accordance with BS1377 (1975) (Test No. 15) on material which is suspected of not being compacted adequately, and compared with tests made on adjacent approved work may show that further compaction is required in the area.

Another factor which may affect the strength of the sub-grade is the level of the water table. An increase in moisture content can reduce the strength and bearing capacity of the sub-grade.

Table 8.2   Table of CBR values for soil found in UK

Type of soil	Plasticity index (per cent)	CBR (per cent)	
		Depth of water-table below	
		More than 600 mm	600 mm or less
Heavy clay	70	2	1*
	60	2	1.5*
	50	2.5	2
	40	3	2
Silty clay	30	5	3
Sandy clay	20	6	4
	10	7	5
Silt	–	2	1*
Sand (poorly graded)	non-plastic	20	10
Sand (well graded)	non-plastic	40	15
Well-graded sandy gravel	non-plastic	60	20

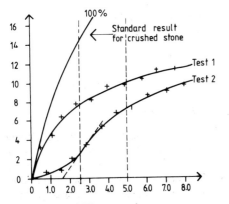

Figure 8.14   Typical CBR test results

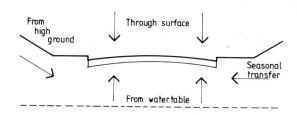

Figure 8.15 Ways in which water can enter and leave the subgrade

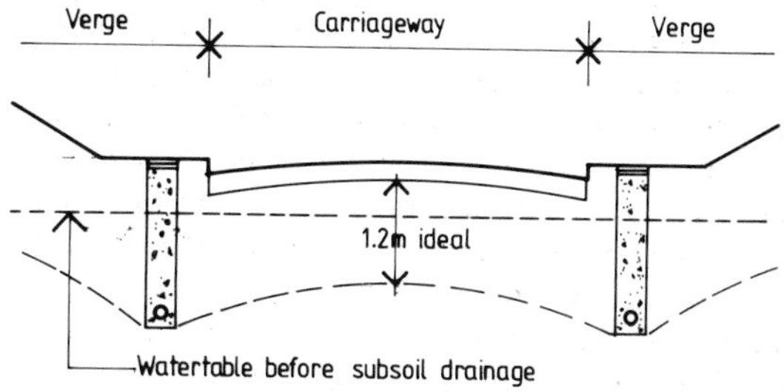

Figure 8.16 Effects of subsoil drainage on water table

Road Note 29 distinguishes between a water table within 600 mm of the formation and a water table more than 600 mm below formation. Ideally the water table should be a minimum of 1.2 m below formation, but moisture movement due to seasonal changes can take place in a number of ways, see Figure 8.15.

To reduce such changes and maintain a stable moisture content in the sub-grade where the need arises, sub-soil drains are provided parallel to the pavement, and on either side as shown in Figure 8.16.

Preparation and surface treatment of the formation should only be carried out after completion of the sub-grade drainage and immediately prior to laying the sub-base or road base material. This treatment should take the form of reinstatement of any soft areas in the sub-grade, and the removal of mud and slurry. The surface should be compacted with a smooth wheeled roller and the formation trimmed to shape and rolled with one pass of a smooth wheeled roller.

Where the formation is not to be protected immediately by the sub-base or road base, it should be covered with a plastic sheeting to prohibit the ingress of water.

To protect the formation before final shaping it is usual to remove material to the full width of the carriageway, and to within 300 mm of formation level. The final trimming to formation level is carried out in one operation and construction traffic should be prohibited on the formation.

When the sub-grade has been prepared the sub-base should be constructed as quickly as possible and only essential traffic should be allowed on the formation prior to the pavement being constructed.

The function of the sub-base is as a structural layer which will bear greater stresses than the sub-grade, and, in the case of a sub-grade which may be frost susceptible, it has a role of acting as an insulating layer as well as providing a temporary road which the construction vehicles may use. Sub-base material is usually brought onto site by side or end tipping lorries which discharge their loads onto the sub-grade. The material

is then spread using a motor grader to give a uniform thickness, and rolled in the manner specified in Figure 8.7 except for lean concrete and dry bound macadam. Where the sub-base or road base are unbound materials, the top 75 mm should be scarified, re-shaped and re-compacted to obtain compliance with the specification for profile and tolerance.

The Department of Transport Specification for Road and Bridge Works defines two unbound granular sub-bases both of which when compacted give adequate strength requirements for pavement design.

Type 1 materials for sub-bases are crushed concrete, crushed slag, crushed rock or well burnt plastic shales.

Type 2 materials for sub-bases are natural sands and gravels.

Where type 1 and type 2 material is not readily available, cement bound materials may be used if it is economical to do so. They are divided into soil cement, cement bound granular material and lean concrete, and since they may also be used as a road-base material, they will be described under those materials.

The road base is the principal load carrying layer, which supports the surfacing and reduces stresses on the sub-base and sub-grade. The Department of Transport specifies three groups of materials:

(i) Unbound material
(ii) Cement bound material
(iii) Bituminous material.

(i) Unbound bases are specified as

(a) Wet-mix macadam

A crushed rock or slag material of specified grading with a sufficient moisture content to give maximum compaction as determined by tests carried out to BS1377. The material is spread in layers not exceeding 200 mm thick and compacted in accordance with Table 8.3.

Table 8.3 Compaction plant and maximum depth of layers (*From Specification for Roads and Bridge Works*)

Compaction requirements		D = Maximum depth of compacted layer (mm) N = Minimum number of passes					
Type of compaction plant	*Category*	*Cohesive soils*		*Well graded granular and dry cohesive soils*		*Uniformly graded material*	
		D	N	D	N	D	N
Smooth wheeled roller	Mass per metre width of roll: Over 2100; up to 2700 kg Over 2700; up to 5400 kg Over 5400 kg	125 125 150	6 6 4	125 125 150	10 8 8	125 125 Unsuitable	10* 8*
Grid roller	Over 2700; up to 5400 kg Over 5400; up to 8000 kg Over 8000 kg	150 150 150	10 8 4	Unsuitable 125 150	12 12	150 Unsuitable Unsuitable	10
Tamping roller	Over 400 kg	225	4	150	12	250	4
Pneumatic-tyred roller	Mass per wheel: Over 1000; up to 1500 kg Over 1500; up to 2000 kg Over 2000; up to 2500 kg Over 2500; up to 4000 kg Over 4000; up to 6000 kg Over 6000; up to 8000 kg Over 8000; up to 12000 kg Over 12000 kg	125 150 175 225 300 350 400 450	6 5 4 4 4 4 4 4	Unsuitable Unsuitable 125 125 125 150 150 175	12 10 10 8 8 6	150 Unsuitable Unsuitable Unsuitable Unsuitable Unsuitable Unsuitable Unsuitable	10*
Vibrating roller	Mass per metre width of a vibrating roll: Over 270; up to 450 kg Over 450; up to 700 kg Over 700; up to 1300 kg Over 1300; up to 1800 kg Over 1800; up to 2300 kg Over 2300; up to 2900 kg Over 2900; up to 3600 kg Over 3600; up to 4300 kg Over 4300; up to 5000 kg Over 5000 kg	Unsuitable Unsuitable 100 125 150 175 200 225 250 275	12 8 4 4 4 4 4 4	75 75 125 150 150 175 200 225 250 275	16 12 12 8 4 4 4 4 4 4	150 150 150 200 225 250 275 300 300 300	16 12 6 10* 12* 10* 8* 8* 6* 4*
Vibrating-plate compactor	Mass per unit area of base plate: Over 880; up to 1100 kg Over 1100; up to 1200 kg Over 1200; up to 1400 kg Over 1400; up to 1800 kg Over 1800; up to 2100 kg Over 2100 kg	Unsuitable Unsuitable Unsuitable 100 150 200	6 6 6	Unsuitable 75 75 125 150 200	10 6 6 5 5	75 100 150 150 200 250	6 6 6 4 4 4
Vibro-tamper	Mass: Over 50; up to 65 kg Over 65; up to 75 kg Over 75 kg	100 125 200	3 3 3	100 125 150	3 3 3	150 200 225	3 3 3
Power rammer	Mass: 100; up to 500 kg Over 500 kg	150 275	4 8	150 275	6 12	Unsuitable Unsuitable	
Dropping-weight compactor	Mass of rammer over 500 kg Height of drop: Over 1; up to 2 m Over 2 m	600 600	4 2	600 600	8 4	450 Unsuitable	8

For items marked * the roller shall be towed by a track laying vehicle.

(b) Dry bound macadam

Is a layer of single size crushed slag or rock of 50 or 40 mm nominal size which is uniformly laid to a thickness of between 75 and 100 mm and given two passes of a smooth wheeled roller. Fine aggregate is then spread on it to a thickness of 25 mm and vibrated into the voids by a vibrating roller or vibrating plate compactor. This operation is repeated until no more will penetrate into the surface. It is then brushed to remove excess fine material and the whole operation repeated until the full specified thickness is reached.

(c) Crusher run road bases

Depend for stability on the particle interlock of the material. These materials cannot be closely controlled for grading and an attempt is made to ensure the presence of particles of various sizes from 75 mm maximum size down. It is brought to site and compacted in layers of 100–150 mm available in some districts, and therefore relatively cheap, it is specified by the authorities, but because of the low moisture content when the material is brought to site maximum compaction cannot be achieved and some deformation under traffic conditions after the road is opened may be experienced.

(d) Colliery shale road bases

Have been used for lightly trafficked roads. The material must be well burnt, not likely to soften in water and not be frost susceptible. Accumulation of these waste materials often leads to them being available at low cost and there is sometimes pressure for them to be used in road construction, but it must be borne in mind that they should only be used if it is certain that they will fulfil their function satisfactorily and not be affected by moisture or frost action.

(e) Hardcore

If used in road construction is usually specified as clean rock-like material such as broken concrete and sound broken brick free from mortar, plaster and wood. Hardcore has been used on factory roads and estate roads, it is laid on a granular material which it can bed into and it needs heavy rolling, a minimum thickness of 150 mm should be specified and any 'hungry' areas are blinded with a fine hard material to ensure a closed surface finish.

(ii) Cement Bound Bases are specified as:

(a) Soil cement

This process requires the complete mixing of the soil with cement to give an average crushing strength of 2.8 MN/m^2 after seven days curing on a 150 mm cube.

Basically the cement is used as a binder to strengthen the soil, to do this a more careful control must be exercised over the specification and works than on other cement bound bases. The 'mix in place' method involves:

Pulverising the soil to a depth of 200 mm;
Spreading a uniform layer of cement to give the required strength after compaction;
Adding water as necessary to meet the compaction requirements;
Mixing together the soil-cement and water to the full depth;
Compacting with a suitable roller.

A large variety of plant is available for this process from agricultural machines to purpose built plant. As an alternative to the 'mix in place' method the soil may be taken to a 'stationary plant' and mixed with the cement in paddle or pan type mixers, brought back to site, laid by a bituminous type paver and compacted.

(b) Cement bound granular material

Is a granular material mixed with cement. The aggregate can be a naturally occurring gravel-sand, a washed or processed granular material, crushed rock or slag or any combination of these, and be sufficiently well graded to give a well closed surface finish after compaction in accordance with Table 8.4.

The cement content must be sufficient to give an average crushing strength based on 150 mm cube of 3.5 MN/m^2. This material, like cement soil, is mixed in pan or paddle mixers which are considered more efficient for this type of material.

(c) Lean concrete

Can be produced in ordinary free fall concrete mixing plants as well as paddle type mixers. The aggregate to cement ratio is usually between 1:15 and 1:20 and the average 28 day crushing strength of a group of three 150 mm cubes should be such that not more than one in any consecutive five such averages is less than 10 MN/m^2 or more than 20 MN/m^2. The aggregate consists of either coarse and fine aggregate batched separately or an all-in aggregate with a maximum nominal size not exceeding 40 mm.

When the material has been mixed it is carried to the site in vehicles suitable for the task. During transport and while awaiting tipping it is protected from the weather by sheeting the vehicles, the material is usually laid by bituminous pavers adapted for the purpose but on many sites it is spread by grader or angle dozer, raked to profile and rolled by a smooth-wheeled roller to attain the required degree of compaction.

(iii) Bituminous road bases

The fifth edition of the Department of Transport Specification for Road and Bridge Works (1976) lists three types of coated material for use in sub-bases.

(a) Dense tarmacadam road base

This material is to comply with the general requirements of BS4987 (1973), there are particular binder requirements to be observed when using flint gravel aggregates and a minimum layer thickness of 60 mm may be used to achieve the specified road base thickness.

(b) Dense bitumen macadam road base

This material must also comply with BS4987 (1973) where gravel other than limestone gravel is the aggregate. The material passing the 75 mm sieve should include Portland cement or hydrated lime.

(c) Rolled asphalt road base to BS594 (1973)

Rolled asphalts are designed in a different manner from coated macadams. The binder normally used for rolled asphalt road base is petroleum bitumen of penetration 50 (50 pen), the mixing temperature of the dried aggregates is between 150°C and 205°C and the binder temperature at mixing is up to 175°C.

On discharge from the mixer the temperature should not exceed 190°C and at the site of laying the temperature should be between 125 and 190°C. All these temperatures are much higher than those for dense bitumen and tarmacadams and this reflects the harder binder used in the asphalt.

These coated materials should be laid by a bituminous paver in layers to give a maximum compacted thickness of 100 mm. Compaction is by means of an 8–10 tonne smooth wheel roller or by a pneumatic roller of equivalent weight. A feature of dense bituminous road bases is their load spreading qualities as compared to uncoated materials, a further advantage may be derived from an early application of the bituminous materials since this will protect the sub-base and sub-grade from the ingress of water.

Bituminous surfacing materials

The functions of bituminous surfacings are:

(i)   To prevent the ingress of water
(ii) Resist deformation due to traffic stress
(iii) Provide a highly skid-resistant surface
(iv) Provide a satisfactory riding surface

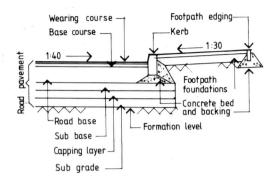

Figure 8.17   Typical flexible construction detail

Surfacing can be laid as a single course but it is usually laid as two separate elements of base course and wearing course – see Figure 8.17, the base course being about 60% of the total thickness of surfacing.

Rolled asphalts to BS594 (1973)

Rolled asphalt is made by mixing a bitumen of low penetration with a graded crushed rock, slag or gravel aggregate together with crushed or natural sand and a filler such as limestone dust or Portland cement.

Wearing course mixtures vary in coarse aggregate content from 30% to 35% depending on the intensity of traffic expected, base course mixtures have a much higher coarse aggregate content of about 60% with a lower percentage of binder and little or no filler. Coated chippings of 14 or 20 mm nominal size are rolled into the surface as part of the laying operation to give a skid resistant surface.

Dense coated macadam to BS4987 (1973)

These materials can be made with tar or bitumen binder, they are almost impervious to water but may need to be surface dressed to give a satisfactory resistance to skidding. The aggregate for the wearing course is usually 10 or 14 mm and for both basecourse and wearing course the aggregate may be crushed rock, gravel or slag with a filler which is usually ground limestone dust.

The material is laid by a paver or by hand in areas such as footways or in regulating courses, but in these cases the binder viscosity is normally lower to allow the material to be handled and adequately compacted.

Open textured coated macadams to BS4987 (1973)

A feature of these types of material which are used for both base course and wearing course is that the material has a low fines content not exceeding 25% which, while making them easy to lay, results in an

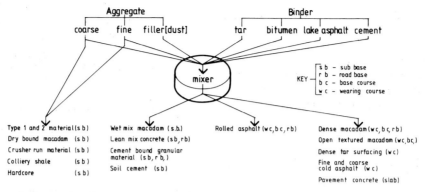

Figure 8.18 Roadmaking materials

open textured surface. The binders are in general less viscous than those used for dense macadams and this permits them to close up under the action of traffic thereby becoming stronger and less permeable. In the main they are used as a surfacing material on lightly trafficked roads.

Cold asphalt surfacing

Fine and coarse cold asphalts are used to provide a wearing couse on footways and lightly trafficked roads. The material is laid to a thickness not greater than 20-25 mm and compacted with a smooth wheeled roller. The name implies these materials are cold but in fact for surfacing purposes they are laid warm.

Dense tar surfacing (DTS) to BS5283 (1975)

Dense tar surfacing is impervious to water and is not greatly affected by the spillage of diesel from vehicles. It is a hot process material consisting of a mixture of coarse and fine aggregate, filler and a high viscosity tar. The material can be used for surfacing new roads with a medium traffic flow, and for resurfacing existing carriageways carrying normal commercial vehicles of up to 2000 per day (sum in both directions). The material should be laid by paver, but on small sites, it can be laid by hand.

Pavement quality concrete

The concrete pavement slab thickness will depend on the expected traffic intensities over the design life of the road and, to some extent, on the bearing capacity of the sub-grade. As previously stated, the methods of estimating this for design purposes can be found by the CBR test. Pavement quality concrete must meet the Department of Transport specification requirements for cement content strength, and air entrainment where required.

The relationship of the materials and their position in the road pavement is shown in Figure 8.18.

Flexible pavements rely on the load spreading properties of a layered system of construction to distribute the wheel loads, imposed on the pavement surface, over the sub-grade. A rigid pavement consists of a concrete slab, resting on a relatively thin sub-base, which acts like a plate to distribute the wheel loads over a much wider area of sub-grade.

Because of these two essentially different methods of distributing the stresses applied to the pavement surface, over the sub-grade, different methods of construction are employed.

Flexible construction

Two preliminary requirements to the satisfactory laying and compaction of coated materials are that:

(a) They should be delivered to site at a suitable temperature.
(b) The surface on which they are to be laid should be structurally sound and of the correct profile.

With the exception of mastic asphalt all pre-mixed bituminous materials for pavement construction may be laid by machine although hand laying can be carried out on small sites, and in awkward areas. The material is delivered to the site in lorries with insulated tipping bodies covered by tarpaulins, and it should remain sheeted until required.

The bituminous material is fed from the lorry into a mechanical spreader and finisher capable of laying to the required widths, profile, camber or crossfall without causing segregation or dragging of the material – see Figure 8.19.

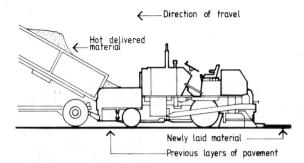

Figure 8.19 Flexible paving machine

It should be operated at a speed consistent with the character of the material mix and thickness being laid, so as to produce a uniform density and surface texture. The material should be fed into the spreader at such a rate as to permit continuous laying as far as site conditions will permit. The material should be rolled in a longitudinal direction as soon after laying as possible but without causing undue displacement of the material. The rolling should continue until all the roll marks on the surface have disappeared.

In the case of rolled asphalt, wearing courses containing 40% or less or coarse aggregate, 14 or 20 mm pre-coated chippings are spread onto the surface of the asphalt prior to rolling. They should be rolled into the surface while it is still plastic enough to cause some embedment of the chippings, this gives the running surface a roughened skid resistant texture.

Rigid construction

Once the sub-base has been prepared to the required tolerance the construction of the concrete slab can proceed. It is normal practice to use a separation membrane immediately under the slab to act as a slip layer, this membrane is invariably polythene sheeting.

In alternate bay construction used on smaller sites, side forms are set up for a number of alternate bays, leaving the intermediate slabs to be cased once the first bays have cured and hardened. The maximum bay length is governed by the joint spacing and would be approximately 13 m. Stop ends are placed in position at the correct joint spacing and the expansion and contraction joints with load transfer bars through them are set up prior to concreting – see Figure 8.20.

Concrete is spread by hand over the slab and the reinforcement mesh is placed in position when the concrete has reached the required level. The top surface of the slab is compacted using a vibrating tamping beam.

To enable tamping to be carried out the side forms are set to the finished surface level.

On larger road schemes mechanised construction is employed and there are two basic methods in use in this country – slip form and fixed form paving.

In fixed form paving the spreading, compaction, finishing and curing of the concrete is carried out between side forms. These are normally steel and fixed to the sub-base by road pins, occasionally these side supports are a part of the permanent construction and take the form of concrete edge strips constructed in advance of the paving. Besides supporting the wet concrete these side forms also carry the rails on which the individual machines making up the concrete train run – see Figure 8.21.

Slip form paving in which the plant has its own travelling side forms is the second method of concrete paving used in this country.

It differs from the conventional concrete train in that

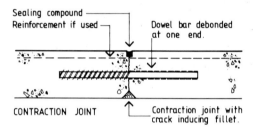

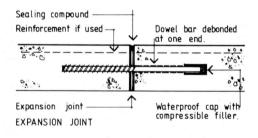

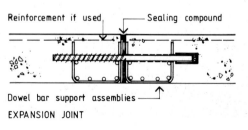

Figure 8.20 Types of joint

Figure 8.21 Concrete paving train

Figure 8.22 Slip form paver

all the operations of spreading, compaction and finishing are carried out within the length of a single machine. The travelling side forms provide edge support to the wet concrete only during the concreting operation. Therefore control on the concrete workability is essential. The machine is guided for line and level by sensors following wires set out accurately ahead of the paving operations – see Figure 8.22.

In both methods of mechanised pavement construction the expansion and contraction joints are formed either in advance of the paving or by mechanical placing during the paving operation.

The purpose of the joint is to accommodate movement either arising from initial shrinkage or thermal movement of the slab or where the slab abuts an existing carriageway. Load transfer is achieved by means of mild steel dowel bars passing through the transverse joint.

Index